MW00965411

THE LEGACY OF HEAVY METAL POISONING

By

Kenneth Frederick Ashton

PublishAmerica
Baltimore

First printing

PublishAmerica has allowed this work to remain exactly as the author intended, verbatim, without editorial input.

ISBN: 978-1-4489-4781-2
PUBLISHED BY PUBLISHAMERICA, LLLP
www.publishamerica.com
Baltimore

Printed in the United States of America

The Legacy of Heavy Metal Poisoning

by Kenneth F. Ashton

The history of toxic exposure from contact with lead, mercury and other heavy metals has been traced to ancient times. Lead is a poison which accumulates in the body; it is dangerous in any amount. It is not excreted, but is stored in the body tissue and bone, then released in the bloodstream. Research demonstrating this has been done over the years by medical practitioners and scientists.

The population involved in this study includes those persons having any contact with the element lead, or other heavy metals, by ingestion or inhalation.

The objectives of this study are to alert professional and lay persons to the danger of heavy metal poisoning, to provide a solution to the problem and include the reasons why the problem has occurred, and to give accounts of cases of poisoning that have occurred. Research will include testing of lead-contaminated materials and testing of radioactive pottery.

There is not sufficient awareness of the danger of heavy metal poisoning. Lead is ubiquitous; it will continue to poison the populace.

TABLE OF CONTENTS

CHAPTER I
Proposal

A. Problem

The hazard of heavy metal poisoning is not one which has become of concern because of recent technological developments. The history of toxic exposure from contact with lead, mercury and other heavy metals has been traced to ancient times. The use of lead began over eight thousand years ago in the Aegean Sea. Lead beads have been found at Cat Huzuk in Asia Minor; these beads dated back to 600 B.C. A small fire provided temperatures sufficient to melt and extract lead from galena, which contains lead sulfide.

The Romans used lead lined vessels in wine making to boil grape juice for flavor enhancement and also to sweeten the wine. One of the compounds of lead is commonly called sugar of lead. Hippocrates commented on the dangers of metallurgical occupations. Slaves forced to labor in the silver mines of ancient Greece rarely survived over five years. Four hundred (400) parts of lead were produced as a byproduct of each part of smelted silver, so lead became readily available for use in plumbing, cooking, and for lining wine storage vessels for those who could afford it.

B. Objectives

Objective 1: To alert professional and lay persons to the danger of heavy metal poisoning and to provide guidelines for persons to avoid being poisoned by heavy metals, lead in particular.

7

Objective 2: To provide information regarding a method of testing for lead and other heavy metals in pottery and painted surfaces that can be used in the home or in industry.

Objective 3: To encourage others to carry on research regarding the danger of heavy metal.

C. Why Is This Information Needed?

Lead is a poison which accumulates in the body; it is dangerous in any amount. It is not excreted and is, instead, stored in the body tissue and bone, then released in the bloodstream. The effects of lead in the body are insidious (more dangerous than evident) and can range from headaches and pains in the joints, to personality changes and decreasing I.Q. Especially vulnerable are middle-aged men, pregnant women and the unborn child, as well as children between five months old and six years of age.

D. Limitations of Study

This study is not intended as a substitute for medical advice regarding particular ailments or diseases. Medical advice should be sought in every case where there is a suspicion of heavy metal poisoning.

NOTE: Definitions of technical, scientific and medical terms can be found at the end of each chapter.

E. Lead Poisoning—Symptoms

Symptoms of lead poisoning are diminished mental and physical abilities and high blood pressure. Personality changes include aggression, irritability, impatience, weight loss, nausea, headache, severe abdominal cramping, gastrointestinal dysfunction, constipation, anemia, nervous system disturbances, dehydration, flu symptoms, insomnia, nephritis, renal diseases, gall bladder problems, and carpal tunnel syndrome symptoms. Some symptoms caused by lead poisoning mimic those of other diseases. This may cause medical

8

practitioners to misdiagnose a disease—for example, carpal tunnel syndrome and acute intermittent porphyria.

Symptoms in children include premature birth, low birth weight, long-term behavioral and learning problems, hearing loss, developmental delays, hyperactivity, mental retardation, blood formation impairment, metabolism impairment, nerve damage, severe poisoning causing kidney damage, brain damage, irreversible retardation, convulsion, coma, and even death.

CHAPTER II
Review of Literature

Wedeen, Richard P. *Poison in the Pot: The Legacy of Lead*. (1984) Southern Illinois University Press.

The belief that modern technology has eliminated lead poisoning is a medical myth abetted by industry, contends Dr. Richard P. Wedeen in his lively historical account, proving that what you don't know can kill you. Wedeen found evidence that workers in the lead industry and many urban dwellers were subject to kidney damage induced by lead.

The insidious scourge of lead from antiquity through the 20th century is the subject of Wedeen's historical examination of lead poisoning and its impact on political, social, and medical life. Wedeen deals with historical as well as contemporary evidence surrounding the effects of lead poisoning. Examining the history and the political implications of contemporary problems with lead, Wedeen concludes that "in modern times workmen have replaced the Roman aristocracy as the major victims of unrecognized plumbism (lead poisoning)." Wedeen contends that "the fruit of man's labor has displaced the fruit of the vine as the focus of concern."

Physicians remain key combatants on both sides of the fray, but their attention has shifted from colic, palsy, and the gout to the delayed effects on the brain and kidney. Wedeen states that it is these organs which appear to be the critical targets of chronic low-level exposure. *Mental Retardation in School and Society*, by Donald L. MacMillan. Boston: Little, Brown, and Company, 1977.

Donald McMillan's book achieves the ideal balance by presenting a broad range of material drawn from many disciplines but with a decided approach to mental retardation as a psychosocial and educational challenge. MacMillan has comprehen-sively treated the salient issues facing the field. His authoritative and well-documented discussions of topics such as the effects of labeling children, testing minority children, mainstreaming, due process, diagnostic-prescriptive techniques, and the impact of mental retardation on the family make this a most significant and useful introductory text.

MacMillan shows that concern for mentally retarded persons has generated a vast amount of professional activity among the major disciplines within the biological, behavioral, and social sciences. In chapter four, he gives an account of a nine-year-old child (Mike) with a tested I.Q. of 64, who was referred to a special education place-ment committee after he repeated two grades on top of a history of academic failures. Some of the information that the committee had presented as background material demonstrates the complexity of forces that could have caused Mike's limitations in intellectual functioning.

Mike is the fourth of seven children who live in a three-bedroom tenement apartment with their parents. The family lived in an extremely low socioeconomic status neighborhood, where the proportion of residents on government assisted programs is the highest in the city. When she was pregnant with Mike, his mother did not receive any medical attention; she did not go to a doctor, and the delivery was performed by a neighbor. Since Mike's father was unemployed at the time, her nutritional intake was very poor. During the fourth month of the pregnancy she hemorrhaged for several days, but no medical help was sought.

As a young child, Mike had his share of illnesses. When he had roseola (roseola infantum, a benign viral endemic illness characterized by a high fever), his temperature rose to 105 F and stayed there for three days. As another medical hazard, the paint peeling from the walls and ceiling of the family's apartment was lead based, subjecting Mike's body to the ravages of lead poisoning.

McMillan, in his chapter on genetic and physical factors, explains that heavy metals are particularly damaging to the nervous system. Lead, arsenic, and mercury are among the most common metals involved in retardation and learning disabilities. Lead ingestion has been a recognized pediatric problem for over 50 years. Close to 50% of the weight of paints used in lower-income housing was lead. As these dwellings deteriorate, flaking and peeling of paint is common, allowing infants and young children to inhale and ingest sizable quantities of lead-based paint. Many of the symptoms of early lead poisoning are not dramatic. The individual may become irritable, constipated, pale, and tired from anemia. With increased ingestion of lead, vomiting and convulsions occur. Repeated exposure is expected because the lead remains available.

In his summary McMillan considers a wide variety of genetic and environmental influences impinging on the developing human being. He concludes by commenting on the placement of the severely retarded in government-based facilities.

Anatomy and Physiology (2nd ed.), by Rod Seeley and Trent Stephens (St. Louis: Mosby, 1992).

With a B.S. in zoology from Idaho State University and an M.S. and Ph.D. from Utah State University, Rod Seeley has built a solid reputation as a widely published author of journal and feature articles, a popular public lecturer, and an award-winning instructor. Very much involved in the methods and mechanisms that help students learn, he has in this text contributed his teaching expertise and proven ability to communicate effectively in any medium.

Professor Trent Stephens has in this text used his versatility as an educator in human anatomy, neuroanatomy, and embryology; his skill as a biological illustrator has greatly influenced every illustration in the text. With a B.S. and M.S. in zoology from Brigham Young University and a Ph.D. in anatomy from University of Pennsylvania, Professor Stephens uses his background to show that human anatomy and physiology comprise the study of how the human body is organized and how it functions. Knowledge from such study makes it possible to predict how a cell-organ system will respond to various stimuli and how this response will affect the whole person.

Since the study of anatomy and physiology is an essential prerequisite for those planning to pursue the health sciences, this text provides a firm knowledge of the body's structure and function, which is essential for health professionals to perform their duties adequately.

Mineral Facts and Problems, by the Bureau of Mines, U.S. Department of the Interior. Washington, DC: United States Printing Office, 1960.

This publication was prepared in response to widespread public interest resulting from the publication of previous editions. Information to prepare this bulletin came from a wide variety of sources. Much of the information from previous editions has been revised as a result of reviews by authorities outside the Bureau of Mines. Many constructive comments had been received.

The Bureau of Mines schedules its research under a formal system having two basic requirements: 1) current work, reviewed and evaluated frequently and 2) new projects, undertaken only after careful appraisal has shown that they will contribute towards solving specific problems, which are defined in program statements prepared for each mineral commodity. The statements are revised periodically, and they reflect the rapidly changing positions of separate commodities in relation to national demands. Thus, they provide information from which ideas for research can be generated and evaluated.

In fulfilling its role of government responsibility for raw material supplies and their efficient utilization, the Bureau of Mines in this publication has analyzed the factor in current and future mineral supplies and demands and has developed statements for specific commodities that assess the national position and problems for these metals, minerals, and fuels. These statements are based on the assumption that nearly all production and consumption of materials in the United States are to be by private individuals and organizations, in accordance with the nation's devotion to a private-enterprise economy.

The objectives of this publication are to encourage and develop: 1) wise production and utilization of United States mineral resources; 2) discovery and development of new resources of mineral supply; 3) maintenance of mineral reserves and stocks at adequate levels; and 4) fostering a productive and processing industrial capacity large and flexible enough to exploit effectively the domestic mineral resources with full foreseen requirements.

Furthermore, the objective of the Bureau of Mines is to be found in the following five points: 1) appraisal of mineral position; 2) development of submarginal resources; 3) possible new or wider uses for abundant resources; 4) development of substitutes; and 5) conservation and wise use of minerals.

The Industrial Environment—Its Evaluation and Control, by the National Institute for Occupational Safety and Health, Public Health Service Center for Disease Control, U.S. Department of Health, Education, and Welfare (Washington, DC: Superintendent of Documents, U.S. Printing Office, 1973).

This publication has become an industrial standard hygiene textbook rather than syllabus. The subject matter is extremely broad, covering subjects from mathematics to medicine. The first two chapters, in addition to providing historical information, cover such areas as mathematics, chemistry, biochemistry, physiology, and toxicology. Other chapters deal with areas of interest to those concerned with evaluating the potential harmful effects of physical and chemical air contaminants. Chapters on safety, solid waste, and water pollution have been added to the previous version.

General principles in evaluating the occupational environment require a multidisciplinary approach. This point is emphasized in the introduction to the publication. A fundamental need exists for the input of the knowledge of engineers, health physicists, physicians,

toxicologists, nurses, and production supervisors in the effort to eliminate hazards which threaten workers. Emphasis is placed on the coordination of many disciplines and effective communication between the employer and the employee for the recognition, evaluation, and control of potential hazards.
General principles mentioned to control potential hazards are listed:

1. Recognition of potential hazard.
2. The investigator must become familiar with all processes used in the particular plant. It is of the utmost importance to obtain a list of all chemicals used in the plant or establishment.
3. Toxicity of raw materials and products. After the list of chemicals is obtained, it is necessary to determine which of these are toxic and to what degree.
4. Selection of instruments to evaluate the work environment. Sampling instruments mentioned are generally classified according to type as follows: direct reading; those which remove the contaminant from a measured quantity of air; and those which collect a known quantity of air for subsequent laboratory analysis.

Finally, the publication explains that the ecology picture as a whole is too complex to understand or to control when considered in its entirety. Yet, those who are responsible for the health and well-being of people in the system must keep the total picture in mind.

CHAPTER III
Presentation of Findings

Introduction

A. Overview

Lead poisoning (plumbism, saturnism) is perhaps the most widely studied and described disease known to medicine. The history of toxic exposures from the ingestion of food or drink prepared or served in leaden or lead-glazed vessels has been traced back to ancient times.[1,2] Despite such early knowledge about the dangers of food contact with lead, there is widespread use of earthenware pottery with highly soluble lead glazes in the world today. I believe that this poses an insufficiently recognized hazard to the public health.

Not all earthenware pottery is glazed. The term "earthenware" is correctly used for pottery that has been fired between 1058 and 2174 degrees F, either glazed or unglazed. Most of the world's pottery has been earthenware because of the abundance of clay deposits and the relative ease in reaching the temperatures needed for manufacture.[3] "Glaze," as related to ceramics, is a thin glassy coating, generally containing silica, which is applied to clay pottery by dipping, spraying, or brushing. During firing in a potter's kiln, it fuses and closely resembles a glass surface; thus, the name *glaze*. Properly formulated, applied, and fired, glazes are continuous, impervious, and almost insoluble. Such glazes can seal the surface of pottery to prevent absorption of moisture and liquids and make it easier to clean and more resistant to wear.[4, 5] Generally, glazed pottery fired below 2200 degrees F can be assumed to be lead-glazed (unless stated otherwise); if fired above 2200 degrees F, it can reasonably certain to be lead-free. The currently popular type of pottery referred to as "stoneware" is in the latter category.[5]

Lead is widely used as a component of glazes because it possesses many fine qualities that endear it to the pottery industry. Some of the more important effects of the use of a lead compound on glazes include:[1,3,5]

1. A low melting range and strong fluxing action permit the formulation of durable glazes that mature at relatively lower temperatures than their leadless counterparts.

2. It increases the "stretchability" or modulus of elasticity of the glaze.

3. It reduces the viscosity (increases the fluidity) of the glaze during firing, and in addition, extends this property over a wider firing range.

4. It lowers the surface tension of the molten glaze and contributes to the ability of the glaze to heal over blisters, drying cracks, and other defects, i.e., helps to create surface smoothness.

5. It reduces the tendency of the glaze to devitrify (reduces undesirable surface crystallization).

6. It confers unique color effects and is an excellent solvent for color oxides.

In addition to these factors, the common lead compounds are the cheapest available flux, are relatively insoluble in water, and can be used in a variety of proportions with most glass-forming silicates.[1] In view of these advantageous effects, lead has been used extensively in the production of ceramic glazes for centuries. However, the use of lead also has a very serious drawback. If improperly produced, lead-glazed pottery may leach dangerous amounts of lead into food or drink. The appreciation and control of this hazard in the United States was only recently initiated. In 1971, the Food and Drug Administration (FDA) established a formal compliance program to control lead contamination of food by domestically produced, commercially imported pottery. Other countries, especially the developing nations of the world, are continuing to manufacture large amounts of pottery with soluble lead glazes. The resulting hazard has not been the subject of adequate surveillance, control, and regulation, either in the United States or abroad.

In recent decades the overall problem of careless dispersion of lead in the environment has attracted significant public attention in the United States. The results of mass screening and other studies conducted in the late 1960's and 1970's indicated that undue lead exposure was common among children. The first information on blood lead levels in the population of this country was revealed in a 1982 report of the second National Health and Nutrition Examination Survey (NHANES II—conducted between 1976 and 1980).[7] The NHANES II showed that 1.9 percent of the civilian non-institutionalized population six months through 74 years old had blood lead levels that exceeded the criterion of 30 micrograms per deciliter (ug/dL) established by the Centers for Disease Control in 1978. Among children six months through five years old, the prevalence of elevated levels was 4 percent—or approximately 675,000. The NHANES II associated blood lead levels with specific socioeconomic and demographic variables including age, race, annual family income, and degree of urbanization of the place of residence. The rate of lead-glazed pottery use was not considered. No doubt such use contributed to an increased body burden of lead in a portion of the population. Validation of this concept, however, will require further epidemiologic studies.

The reported frequency of lead poisoning from pottery has been low with only a dozen or so cases of chronic or acute poisoning being documented in the literature during recent years. Extrapolation of such limited data to estimate the overall hazard is not reasonable. However, the combination of a review of some of the literature accounts of lead intoxication from pottery use, the difficulties in diagnosis of plumbism in the non-occupationally exposed, and the abundance of improperly manufactured lead-glazed pottery, give credence to the fact that the hazard is underappreciated and that it may be an important public health problem.

Most studies of plumbism caused by lead-glazed earthenware vessels deal only with a single cup, mug, bowl, or pitcher that was tested for lead release after-the-fact. Results varied widely as to the amount of lead leached, and experimentation usually ceased after a particular vessel was identified as the source of lead in the case in question. While some reporting authorities considered their cases to be isolated or unique, others expressed the opinion that the hazard posed by lead-glazed earthenware is likely to be more common than is recognized.[2, 3, 10-17]

Chronic lead intoxication might easily remain undetected for a long time[2, 7] or be misdiagnosed, especially when there is no reason to suspect that the patient has had abnormal contact with lead.[14, 15] The clinical features of the disease are confusing and much has been written, particularly in the occupational medical journals, about the effects of increased lead absorption without the development of overt clinical symptoms.[16] Additionally, many of the presenting symptoms in adults, such as headache, dizziness, fatigue, malaise, abdominal discomfort, and weight loss are non-specific. There is evidence that such symptoms may occur at relatively low blood lead levels and may be early clinical signals preceding the more acute events of lead poisoning.[17] The nonspecificity of the symptoms makes it difficult to diagnose the disease in young children, particularly in its early phases, when complaints such as stomachache, weakness, irritability, and fatigue may be blamed on teething, viral infections, birth of a sibling, or other stresses.[18] There is a plethora of data in the literature concerning the methods of exposure and the pathology of lead intoxication; the etiology of the disease is clear.

In summary, plumbism occurs when substantially more lead is absorbed—through the air one breathes or from the food or other material one ingests—than one's body can excrete. Lead is highly toxic, i.e., the fatal dose of absorbed lead has been estimated to be 0.5g.[19] From the gastrointestinal and respiratory tracts, lead flows into the blood stream and accumulates in body tissues, particularly the kidneys, bones, and nervous system. Its most serious toxic effects involve the brain and peripheral nervous system. The fetus, infant, and child are particularly vulnerable. Once absorbed, lead remains in the body for months or years;[18] with a half-life estimated to be 32 years in bone and 7 years in the kidney.[18] Estimation of total exposure is difficult and the brain and liver lead levels may be 5 to 10 times the blood level.[19]

Significant lead exposure can be considered to be anything in the milligram per day (mg/day) range. Accumulation and toxicity can occur if more than 0.5 mg/day is absorbed.[19] Kehoe concluded that a rapid increase in body burden of lead, and anemia,[21] can be manifested in the adult with an intake greater than 2.3 mg of lead per day; and that symptomatic lead intoxication can occur in the adult with greater than 5 mg of lead per day for one month. Children are affected with much lower exposures; greater than 300 micrograms (ug) of lead per day can induce a rapid increase in body burden and anemia. More than 1 mg of lead per day for several months can produce symptomatic lead poisoning in a child.[22]

The purpose of this work is to provide information about the potential for dangerous amounts of added dietary lead intake to occur through the use of lead-glazed pottery by demonstrating that lead release may be enhanced during normal household use—particularly from improperly manufactured vessels. For more detailed examination of the spectrum of effects of lead poisoning for clinical recognition, diagnosis, and management, the reader is referred to recent reviews by Dreisbach[19] and Cullen,[20] among others.

B. Lead in Human Foods
1. Historical Perspective

Lead (Pb) is ubiquitous in nature and this natural background level results in its being incorporated into all living organisms.[23] However, it has not been established that lead is an essential nutrient for any organism. The current levels of environmental lead are the result of exploitation of the natural sources by man, which has been ongoing for millennia.[24] Artifacts made from lead that date back some 3,000 years B.C. attest that it is one of the oldest metals known to mankind.[25]

The desire for silver was the principle stimulus for lead production in early times with about 400 parts of lead produced as a byproduct of each part of smelted silver. Settle and Patterson,[26] in citing current estimates, reported that world lead production from 4,000 years ago until about 2,700 years ago averaged 160 tons annually. With the advent of silver coinage, production rose to about 10,000 tons and it rose again to about 80,000 tons during the Roman Republic and Empire. In medieval times lead production declined, but rose dramatically with the onset of the Industrial Revolution—from 100,000 tons annually 300 years ago to 1 million tons 50 years ago. Lead mined and smelted for itself has constituted a significant portion of total lead production only during the past century. Today's worldwide production is about 3 million tons annually.

The toxic effects of lead were known in antiquity with the first reports of lead colic attributed to the time of Hippocrates (400-355 B.C.).[27] Nieander, a celebrated Greek physician and poet of the 3rd century B.C., recognized the disease and its cause.[18] The prevalence of lead poisoning in ancient times is speculated upon in modern literature. Several authorities suggest that the Roman Empire suffered due to lead poisoning among its citizenry. In 1965 Gilfillan[28] suggested that lead poisoning among the Romans was introduced through water collected from troughs and cisterns lined with lead and through wine, grape syrup, and preserved fruits prepared in lead-lined containers. He supported the concept that lead intoxication caused a decrease in

21

reproductive capacity, which may have been a factor in Rome's decline. Neuhauser[29] (1970) attributed the fall to lead-induced infertility in many of the noble families. Also in 1970, Lacambre[30] gave consideration to the arguments that the culture depended mainly on the ruling classes, who began to progressively disappear with each generation, perhaps losing a quarter of its members due to the high infant mortality rate attributed to wide-scale lead poisoning. Nriagu[1, 31] recently reported on the coexistence of wide-spread lead poisoning and gout, a seemingly important feature of aristocratic lifestyles that had not previously been recognized. He considered this to provide strong support for the hypothesis that lead poisoning contributed to the decline of the Roman Empire.

In the 18th century, George Baker, physician to the British Royal Household, recognized that a widespread epidemic of colic was due to lead piping in the apparatus used in making cider. He was widely condemned for his belief, but he managed to convince the cider producers to stop using lead pipes, and the disease disappeared. His recommendation was suggested in correspondence with Benjamin Franklin. Franklin had mentioned that the use of lead had been prohibited in equipment used for making alcoholic beverages in New England as a result of health complaints.[18, 32]

Widespread episodes of lead poisoning resulting from food contact with lead in its metallic form decreased as the hazard became more widely recognized over time. However, a few notable exceptions occurred; perhaps the most widely known resulted from the use of lead-soldered automobile radiators to distill moonshine whiskey. In December 1983 it was reported that about 300 people in eastern France were being treated for lead poisoning as a result of drinking water contaminated by lead from old pipes.[33]

During recent decades lead contamination in food has been a subject of considerable interest to investigators and regulatory agencies. While various estimates of the sources of lead in the diet have been made, emphasis has been on factors other than the use of lead-glazed pottery.

2. Lead Contamination in Food—Past and Present

Excluding the occupationally exposed, current estimates indicate that, on the average, the general population in the United States receives 70 percent or more of its total exposure to lead through the diet. Water and airborne sources comprise the remainder.[25]

In 1980 Settle and Patterson[26] estimated the natural lead in human foods during prehistoric times—among Peruvians who lived in an unpolluted environment 1,800 years ago—to be less than 2 ng/g (0.002 ppm). They compared this to estimated average concentrations of lead in the modern American diet of about 200 ng/g (0.2 ppm) and noted that the maximum safe intake level recommended by the World Health Organization in 1972 is equivalent to 300 ng/g (0.3 ppm) in the adult diet. This would amount to approximately 430 ug Pb/day from all sources.

In another comparison, the same authors considered absorbed lead. They estimated the total daily absorption of lead into the systemic blood of prehistoric humans to be less than 210 ng/day, with air and water sources accounting for less than 2.3 ng/day of the total. Food accounted for the remainder. Today they estimate a lead uptake of 29,000 ng/day for adult urban Americans (21,000 from food, 6,500 from air, and 1,500 ng from water)—a greater than 100-fold increase. The fraction of lead absorbed from food (7 percent of the total lead in food) was considered to be the same in prehistoric times and today, despite large differences in lead concentrations between the two diets. The calculations indicate that the average urban adult in America today has a lead exposure of about 300 ug/day from food. It should be noted, however, that there is a wide range in literature values for absorption characteristics of ingested lead.[25, 34, 35]

3. Sources of Lead in Food

As mentioned previously, it has been estimated that 70 percent or more of the total non-occupational exposure to lead in the United States is through the diet. Although small quantities of this food-borne lead occur as natural background levels, the majority is the result of environmental pollution and food processing activities.

Environmental pollution sources include lead from automobile exhaust and lead smelting operations, lead in runoff water from mining and industrial dumping, and the use of lead-containing pesticides on certain fruits and vegetables. The most important source of additional lead in the food supply from food processing is the method of packaging and holding food.[23, 25] The FDA estimated that 20-30 percent of the lead in the average daily diet of persons more than one year old is from canned food, of which about two-thirds is from the solder, and the remainder from the food itself prior to canning.[23, 24, 25] Settle and Patterson,[26] however, estimated that half of the lead in the American diet originates from lead-soldered cans.

In August 1979 the FDA announced that within five years it hoped to reduce lead in food from lead-soldered cans by at least 50 percent.[23] The long-range goal of FDA, in conjunction with other federal regulatory agencies, is to reduce lead intake from all sources—air, water, and food—to less than 100 ug/day for children[36] one to five years old.[23] With the progress up to 1982 in reducing lead in foods for infants, the FDA has shifted emphasis to reducing lead in canned foods for adults that are also commonly eaten by infants and children.

The FDA reports that other less significant sources of added lead in food include migration of lead under normal conditions of use from ceramic glazes, silver-plated hollowware, porcelain pots and pans, pewter, and fine leaded crystal. However, the FDA cites that misuse of glazed ceramic ware by storage of acid foods or drinks for prolonged periods can result in the leaching of relatively large amounts of lead into food.[23] The remaining sources of small amounts of lead, according to the FDA, include drinking water and, hence, water used to prepare foods. The maximum permissible level of lead in drinking water is 50 ug/L, although this level may be exceeded in certain localities. In some communities, the problem of elevated lead levels in water is associated with the use of lead plumbing. The FDA also states that very small (trace) amounts of lead in food result from food and color additives; however, these are subject to regulation.[23]

4. Estimated Dietary Intake of Lead

While Settle and Patterson[26] concluded in 1980 that the average urban adult in the United States was exposed to about 300 ug/day of lead from food, more recent estimates indicated the exposure to be considerably lower. In 1982 Jelinek[37] (FDA Bureau of Foods) noted that the FDA has measured lead in its Total Diet (Market Basket) Studies since 1973. The program was initially established in 1965 to identify problems and trends in the dietary intake of chemical contaminants in the food supply. Jelinek stated that the mean daily lead intake for teenage males in the United States varied from about 80 to 90 ug/day for the period 1977 to 1980, well below the tolerable daily intake of about 430 ug/day from all sources proposed by FAO/WHO, 1972.

Biddle[25] (FDA Division of Toxicology) related in 1982 that numerous studies have been conducted to quantify dietary lead intakes during the past 30-40 years. He noted that considerable variation can exist, depending upon a number of factors including geographic location, agronomic practice, food preferences, and socio-economic conditions. He provided a brief summary of the estimates for dietary lead intake by infants, toddlers, and adults for the period 1975-1980, based upon the "market basket" concept of the Total Diet Studies program. The adult market basket consisted of about 120 food items representing the two-week diet of a teenage male. The summary is shown in Table 1.

Table 1
Mean Daily Dietary Lead Intakes as Determined
in FDA Total Diet Studies Program

Fiscal Year Dietary Lead, ug/day

	Infant	Toddler	Adult
1975	21	36	67
1976	27	30	71
1977	22	28	79
1978	25	35	95
1979	36	46	82
1980	34	43	83
PTDI[a]	-	-	429
	100[b]	150[b]	-

[a] Provisionally tolerable daily intake for the adult from all sources, established by FAO/WHO, 1972.

[b] Maximum intake from all sources recommended by FDA.

The estimations for the adult dietary lead intake per day reflected in Table 1 are similar to those presented by Jelinek (80-90 ug/day); however, both are considerably lower than the 200 ug/day noted by Settle and Patterson.[26]

In 1983 Bander et al.[38] published the results of a nationwide, seven-day food consumption survey of 371 preschool children between the ages of birth and five years. Results indicated that the average daily dietary lead intake was 62 ug Pb/day (range 15-234 ug/day). The greatest percentage of children had a mean intake of between 45 and 57 ug Pb/day. The authors reported that their results revealed average daily dietary lead intake to be lower than that found by previous researchers. Further, they concluded that the majority of sampled children's daily lead intake from food was within the FDA's long-range goal of less than 100 ug for ages one to five.

To determine the average daily dietary lead intake, the FDA used a regional approach. Individual food samples for the adult were collected and prepared as they would be in a household, and combined in 12 consumption weighted food class composites. These were then analyzed for various contaminants, including lead. Canned food was included.

In the study done by Bander,[38] the Michigan State University Nutrient Data Bank, which contains about 3,500 food items, was updated with the lead content of food per 100 g. Average lead values for numerous foods were provided by the National Food Processors Association. The authors stated that these were the most recent data available.

Neither the FDA studies nor the survey conducted by Bander considered the use of glazed pottery as a source of lead. The results, therefore, could be quite low. If a child were to consume 12 ounces of a beverage such as soda pop or juice daily, for example, from a lead glazed cup or mug that release one-fourth ppm of lead into solution, almost 90 ug of lead per day would be added to the diet. Such an example, as will be shown in this work, is not unreasonable. It is believed that millions of lead-glazed vessels in this category are in use in the United States today.

C. Compliance Program—Foreign and Domestic Pottery

The first of the FDA's recent programs concerning lead contamination of food began after the FDA became aware in the late 1960's of a potential problem involving lead migration into food from pottery glazes.[4] There had been wide-spread public concern and reaction in 1969 following the publication of an article describing how a family of five in California had almost died of lead poisoning as a result of their prolonged use of a Mexican handcrafted earthenware pitcher. The pitcher, used as a container for orange juice, caused toxic accumulation in the bodies of a physician, his wife, and their three children.[39] In February 1970 a limited survey (by the FDA) of imported pottery revealed a high incidence of hazardous lead release.

The program was expanded and in fiscal year 1970, the FDA denied entry to over 400 lots of pottery from 19 different countries because of leaching of excessive amounts of lead. In 1971 a formal compliance program was initiated for foreign and domestic pottery to enforce a limit of 7 ppm on lead migration into 4 percent acetic solution over 24 hours. In addition, the FDA inspected factories and analyzed samples from every major domestic manufacturer of ceramic dinnerware. The rate of violation of the 7 ppm limit has been relatively low. The FDA (to August 1979) has also examined approximately 2,000 samples of ceramic ware imported each year and found the violation rate somewhat higher than that of domestic items. Items that exceeded the limit were denied entry into the U.S.[23]

In 1982 the FDA provided information regarding the number of shipments of imported earthenware pottery for the period 1975-1980, the number of examinations conducted for heavy metal content, the number of import denials, and the criteria used for testing. While the data concerning the number of shipments and examinations were not available, the FDA provided detention and criteria information.

Table 2
Number of Detentions of Imported Pottery
Made by the FDA for Fiscal Years 1977-1981

Fiscal Year	Number of Detentions
1977	75
1978	92
1979	48
1980	85
1981	83

It should be noted that data presented in Table 2 refer to detentions of imported lots of pottery, some of which could have occurred due to high cadmium, as well as high lead release, during testing. Nevertheless, the number of recent detentions

28

indicate that potentially hazardous pottery continues to be manufactured and exported, possibly making its way to the marketplace in countries that do not have a heavy metal release standard nor import examinations.

The criteria used by the FDA to prevent imported and domestic ceramic kitchenware and tableware of high lead release potential from being marketed in the United States is presented in Table 3. Shipments will be detained by the FDA if spot checking of six units of the categories shown meets the action level for lead release when tested.[40]

Table 3
Lead Release Criteria for Domestic and Imported Pottery

Category	Action Basis	Action Level
Flatware[1]	Average of 6 units	7.0 ug/mL
Small Hollowware[2]	Any one of 6 units	5.0 ug/mL
Large Hollowware[3]	Any one of 6 units	2.5 ug/mL

[1] Internal depth not exceeding 25 mm.
[2] Internal depth greater than 25 mm; capacity of less than 1.1 liter.
[3] Internal depth greater than 25 mm; capacity of 1.1 liter or more.

It is noteworthy that the lead release criterion in Table 3 is 7.0 ug/mL (7.0 ppm) for flatware such as plates and other shallow vessels. Maximum lead release values are less for deeper vessels (over 25 mm internal depth), depending upon vessel capacity. This sliding scale may appear reasonable in that flatware is normally exposed to small quantities of liquid foods, and lead release potential (total quantity of extracted lead) would be less, considering like items excepting size, in smaller vessels than in larger ones. There are, however, deficiencies in the criteria, which will be detailed later. At this point, the following criticisms should be noted:

1. The levels of allowable lead release from various types and sizes of vessels is apparently not well known. Reports often referred only to the 7 ppm maximum, thereby implying that a lesser lead release is acceptable.[11, 42-45]

2. Based upon the toxicity and accumulative properties of lead, the action levels are set much too high. Additionally, items are leached while in new condition with no provision for possible enhancement of lead release after use.

3. Spot checking may not identify the hazard potential because wide item-to-item variation in lead release can exist among like items from the same manufactured lot.

In summary, while the FDA's formal compliance program has undoubtedly reduced the potential for plumbism caused by commercially available pottery (domestic and imported), this progress does not extend to the hazard posed by uncontrolled pottery, i.e., lead-glazed pottery that is not required to be inspected under current FDA guidelines.

D. The Hazard Posed by Uncontrolled Pottery

The term "uncontrolled pottery" was selected to define lead-glazed ceramic dinnerware, both foreign and domestic, that was produced prior to the implementation of the FDA formal compliance program in 1971; pottery produced outside of the United States that has been (and continues to be) imported into this country by individuals without a requirement for testing; and pottery produced and marketed by art potters and hobbyists.

1. Pottery Produced Prior to the FDA Formal Compliance Program

The following summary of surveys and case studies indicates a hazard exists from use of pottery of this category.

The New York Health Department warned in 1981 that bright red-orange dishes made with a uranium and lead-based glaze released lead and a small amount of radio-activity "above current guidelines.[46]"

The glaze was found to flake off badly scratched dishes or when the glaze came into contact with highly acidic foods. The dinnerware in question was manufactured between 1930 and 1970 by several manufacturers in the United States and Europe under brand names including Fiestaware, Caliente, Early California, Harlequin, Poppytrail, and Franciscanware. Some examples have become highly prized collector's items and are widely marketed by antique dealers.

The literature contains accounts of lead-glazed vessels that have been used for decorative purposes for decades and then have expressed their toxic potential after being subjected to food items. For example, a large bowl was over 100 years old when it caused severe lead poisoning in an individual who had inherited it and then used the bowl for making wine. In another case, a pitcher was bought in New York in 1926, and subsequently identified as the source of lead in a poisoning 45 years later in England.[9, 10]

In 1970 Klein[3] investigated a fatal lead poisoning from the use of earthenware pottery and then tested 264 examples of contemporary glaze surfaces for lead release. Leaching was conducted using an exposure to a solution of 4 percent acetic acid for 18 hours. The study concluded that 50 percent of the vessels tested released sufficient lead to make them unsafe for culinary use, and between 10 and 25 percent could have caused severe lead poisoning. The examples included 117 items purchased from handicraft shops and department stores, and 147 items prepared in a ceramics laboratory with use of 49 different glaze combinations. The study was conducted in Canada where, at the time, no standard existed for ceramics intended for culinary use.

2. Pottery Produced by Primitive Methods

Today, lead-glazed earthenware pottery continues to be produced in several countries using unregulated and obsolete methods of manufacture.[44-47] A study conducted by Acra[44] and others during 1979-1980 concluded that the kilns of the primitive potteries in Lebanon (as in other countries in the Middle East) are poorly designed and constructed and operate at uncontrolled temperatures. The raw material used for glazing is a suspension of red lead (Pb_2O_4), which is applied manually as a coating on each item before drying and firing. Glazes produced in this way are not acid-resistant, especially when the firing temperature is not optimum. Of 275 examples of pottery manufactured in this manner and tested using the FDA method,[41] the mean concentration of lead leached was about 97 plus or minus 63 ppm, with a range of 0.2 to 200 ppm. Irregularities in temperature and incorrect stacking of items within the kiln were considered to be some of the factors contributing to the wide range of results. Glazed utensils produced in Beirut by superior methods of manufacture released much lower levels of lead. Of 75 leached, mean lead release was about 3.8 plus or minus 2 ppm, with a range of 0.8 to 9.6 ppm. The authors noted that 91 percent of the examples produced by superior methods met the FDA maximum limit of 7 ppm. They concluded that all pottery with similar glaze quality to that produced by primitive methods in Lebanon, or in other countries of the region, should be condemned as unfit for use as kitchenware and tableware.

It should be noted that the study referenced above employed leaching tests of vessels in new condition. Further, the report did not indicate that additional research was done to determine body burdens of lead in persons using the primitively produced pottery on a regular basis.

In 1980, Tavolato[11] reported on a case involving a woman who developed polyneuropathy from the use of a lead-glazed cup purchased at a local fair in Italy. The description of the cup was similar to Italian earthen-ware examined in this work. The report stressed that the manufacturer of the cup was not identified by enquiring authorities. They deemed it highly probable that other lead-liberating cups of the same type were sold and in use Italy, and possibly in other countries as well.

Clark[12] reported on a case of plumbism that occurred in England in 1972 that was caused by serving cider (pH 3.3) from a reddish-brown earthenware jug that had been purchased in France. He noted that lead-glazed earthenware vessels have been the chief cause of many large epidemics of colic and paralysis, concluding that the public was still not fully aware of the risk. Extraction for four hours with 1.5 liters of 5 percent acetic acid solution (pH 2.4) dissolved the equivalent of 683 mg of elemental lead from the interior of the vessel. Clark believed that since such items are popular tourist souvenirs, the problem of lead poisoning from this traditional source will continue for many years to come.

Goldfield and Altman[48] reported on the case of a 4 year old boy whose blood lead level was found to be elevated (41 ug/dL) during a 1974 blood-lead screening program in New Jersey. The only source of lead in the child's household proved to be pottery the family brought back with them after living in Barbados, West Indies. In testing, a small cup showed 270 ppm, a small bowl 820 ppm and a large casserole dish 2,200 ppm of lead.

In 1977 Koplan[45] reported on a survey conducted in a community of potters in Barbados where lead glazes traditionally have been used. Eight pottery pieces tested by the FDA method released lead in amounts ranging from 7 to over 3,100 ppm (three were over 1,900 ppm). In addition, the survey revealed elevated blood-lead levels in potters who used lead glazes and also in their family members due to exposures to lead-containing dust and to use of the pottery for culinary purposes. Elevated blood-lead levels were also found in controls who were not exposed to dust residue involved with pottery-making. Their exposure was related to use of lead-glazed pottery for cooking and serving food.

3. Pottery Produced by Art Potters and Hobbyists

Demand for handmade pottery has been high with increased production and availability indicated by the popularity of pottery courses.[2, 15] Although only limited data are available, in 1979 Henderson et al.[43] noted many homemade ceramic items leached excessive amounts of lead. They found that none of a total of 71 commercial items (both domestic and imported varieties) released over 7ppm in one hour exposures to a solution of 5 percent acetic acid at 85 degrees C. In comparison, 19 of 33 items of domestic homemade variety released over 7 ppm of lead (11 of which released over 100 ppm) using the same method of extraction.

The major emphasis in this work will be placed upon further defining the hazard posed by uncontrolled pottery, especially lead-glazed pottery produced by primitive methods. Large quantities (perhaps millions) of vessels with soluble lead glazes have been produced using unregulated and obsolete methods. Such production continues today.[44,47] As will be shown, use of this type of earthenware portends a major health hazard, particularly after such vessels are subjected to extended household use.

E. Changes in Lead Release from Glazed Pottery—A Model

Rationale for conducting the investigation summarized in this work was based upon unresolved questions generated after examples of Italian lead-glazed pottery were tested for lead release. Used examples of the earthenware that caused plumbism were found to release very high quantities of lead into various leaching solutions. For example, 0.1 M nitric acid placed in a soup bowl revealed 23,000 ppm of lead; diluted (0.1 M) acetic acid in a coffee cup produced 440 ppm of lead, and hot coffee in a cup produced 80 ppm of lead.

A literature account of a poisoning from lead-glazed pottery resulted in a disagreement regarding the relationship of lead leachability to the history of pottery use. Whitehead and Prior[9] noted that the glaze in a large bowl used in wine-making had become cracked and pitted by the acidity of the wine. Pieces of the glaze were even finding their way into a serving glass. The authors concluded that the lead contamination was not appreciable until after the glaze became cracked and broken. In commenting on the account, however, Beritic and Stahuljak[10] maintained that acid food ingredients kept or boiled in lead-glazed vessels extracted the lead from the glaze until none is left. They concluded that lead contamination would be more appreciable before the glaze became degraded with use.

34

Early in the research phase of this work, leachings of new and used lead-glazed coffee cups from the same manufactured lot of Italian earthenware pottery were conducted. The used cups exhibited a tendency to release much more lead than their counterparts in new condition, i.e., several hundred ppm versus less than 100 ppm. Additionally, the glazed surfaces of the used cups, particularly the interior area that had been in contact with coffee, were less brilliant than those of the cups in new condition.

Coffee had been served in the used cups many times. In addition, the used examples had been subjected to numerous hand and automatic dishwashings. Occasionally, the interior surfaces had been cleansed with a soap pad to remove coffee stains. The used cups were typical examples of several that had been used for about one year during a 2 year time period. A more accurate usage estimate was not possible as the pottery was often placed in storage.

After the initial leaching tests were conducted, scanning electron microscopy (SEM) techniques were employed to examine the glazed food contact areas of both a new and a used cup. SEM revealed that the glaze on the used cup was heavily eroded with apparent degradation throughout its thickness. The glaze on a similar portion of the unused cup appeared smooth and glass-like.

Lead in Pottery and Paint

In the late 1960s, a two year old boy was admitted to the Montreal Children's Hospital in a coma.[49] While in the emergency room, he stopped breathing and was hooked to a respirator, but this and all the other efforts of the medical team proved fruitless. Three days later, the boy died; the diagnosis: death by lead poisoning. The cause: lead leaching from the glaze of an earthenware jug. Three weeks before entering the hospital, the child had been diagnosed as having gastritis, and his physician had recommended rest and forced fluids. The boy was given apple juice, in accordance with the physician's advice, but the juice had absorbed lead from the jug's glaze; it was this lead that caused the boy's death.

Around the same time, a California physician, his wife and three children nearly died from lead poisoning as a result of lead leaching from the glaze of a Mexican handcrafted earthenware pitcher that had been used by the family as a container for orange juice.

These incidents, and earlier cases reported by physicians, prompted both Consumer and Corporate Affairs Canada (CCAC) and the U.S. Food and Drug Administration (FDA) to investigate the danger posed by lead in the glazes on dinnerware. After testing hundreds of samples of dinnerware and related items, CCAC found a high incidence of hazardous lead release. In 1969, a program was initiated in Canada to limit the allowable lead release from all pottery that might come in contact with food. Since then, CCAC has monitored pottery at the import, wholesale and retail levels, requesting that items in violation of the standards be recalled.

The FDA also conducted a limited survey of imported dinnerware following the poisoning incident in California, and in 1970, denied entry to over 400 lots of pottery from 19 different countries. Though exact figures on how much pottery is found to contain lead each year are not available, the FDA has turned up numerous violative lots of foreign pottery going into the United States: between 1977 and 1980, an average of 75 lots were detained each year, and in 1981, there were 83 detentions.

The pottery industry in North America has responded by instituting a policy of evaluating glazes and firing techniques regularly to ensure commercially manufactured ware meets CCAC and FDA standards. Even with these safeguards, however, the danger posed by lead in the glazes of dinnerware has by no means disappeared.

On May 14, 1986, the recall of over 15,000 earthenware casseroles sold by Williams-Sonoma, a major American importer and retailer of cookware, was announced by the FDA. According to the Agency, some of the casseroles, called *Cazuelas,* had been tested and found to be "a potential health hazard for lead poisoning."[49] The tests showed that lead leaching from the glaze on the vessels far exceeded allowable limits. In addition, at least one style of Italian pitcher carried by Pier One, another large American importer with 30 stores in Canada and more than 300 in the United States, was found to be in violation of the standards of both countries. Pier One has pulled the pitchers from the shelves in the United States. CCAC has now alerted its field investigators, and the pitchers in Canada are being tested at the retail level.

The enormous number of dinnerware and tableware items being imported into North America each year make the job of catching every bad piece nearly impossible. Over half a billion pieces were imported in 1983 alone, and even with CCAC and FDA spot checks, the fact is that some of the pottery being sold in North America today, as well as numerous items owned and used by consumers, is in violation of the standards. The Pier One and Williams-Sonoma items, for instance, were discovered to be in violation by an independent laboratory, not by CCAC or the FDA. Though Williams-Sonoma does not sell directly to Canadian consumers, according to a company spokesperson, their products are often brought back by vacationers as gifts or are shipped to the border to be picked up or reshipped into Canada. "We have a network of investigators who periodically check pottery all across the country," says Murray Kaiserman, a scientific project officer with the Product Safety Branch of CCAC. "There are, however, limits to what we can do from a personnel standpoint."[49]

Although domestic pottery produced since 1970 is considered safe by most government officials, industry representatives and ceramics engineers, the lead in imports and early domestic ware still poses a significant risk to public health. Independent investigators, physicians, public health advocates and even some studio potters are now concerned that many people in North America are being slowly poisoned by lead leaching from improperly glazed pottery. In addition, current research indicates a previously unrecognized connection between very small amounts of lead in the diet and several types of chronic disease, and medical researchers are questioning the validity of even CCAC and FDA standards.

This criticism stems from the conflict between limits set by these agencies for allowable lead release from pottery and goals set by several international organizations for maximum elimination of daily lead consumption by individuals. Though no such goals have been formalized in Canada, toxicologists with Health and Welfare Canada are in the process of gathering information in preparation for doing so.

Lead intake goals have been set by agencies such as the World Health Organization, the Food and Agriculture Organization and the FDA. These recommendations state that infants and toddlers should take in no more than 100 micrograms of lead per day, older children no more than 150 and adults no more than 430. Using a one-quart pitcher as an example, the present Canadian standard says that such a pitcher filled with an acidic testing solution can leach up to 7 parts per million (ppm) of lead over 18 hours and still be safe. In a vessel that size a 7 ppm release translates into 7,000 micrograms of lead, or 70 times the stated goal for infants and toddlers from all sources.

The testing solution used by the CCAC is 5 percent acetic acid— the same as household vinegar. There are several acid-containing foods and liquids that might be stored in vessels of this size for 18 hours or longer. White wine, for instance, is nearly as strong as the test solution. Tomato juice, cider and other juices also contain acids, and such things as spaghetti sauce, sauerkraut and salad dressing, which might be served and stored in pottery bowls, are also acidic.

One example often cited is orange juice, which is weaker than the testing solution but which nonetheless could leach up to half as much lead. At that rate, a small pitcher that was allowed to leach up to 7,000 micrograms of lead in 18 hours during testing and still be declared safe could release 3,500 micrograms into the same pitcher filled with orange juice.

No one knows when or where lead glazes were first invented, but according to the book *Lead in Antiquity*, pottery shards with lead-silicate glazes have been found that some archaeologists believe were made as early as 2000 B.C. There are several reasons for using lead in glazes. One is that adding lead allows the potter to create a broad range of bright surface colors that are otherwise difficult or impossible to achieve. Perhaps the main reasons for using lead glaze, however, were economic. Because they are easy to work with, cost little and can be fired at relatively low temperatures, the use of lead glazes requires less technical expertise and less of an investment in initial cost and energy use.

Earthenware, which poses the greatest hazard for lead release, can also be fired at low temperatures, and for this reason, the vast majority of earthen-ware glazes contain lead. For a lead glaze to be safe on food surfaces, it must be properly formulated and then heated in a kiln until the chemical structure of the glaze is rendered virtually impervious. This process, called firing, is crucial to the safety of the finished piece, and improper firing is the primary reason an item may leach dangerous levels of lead.

Because the clay used for earthenware is the most abundant and least expensive of all clays, most of the world's pottery is earthenware. Unlike the clays used for stoneware and fine china, earthenware clays contain mineral impurities that allow them to mature when fired at temperatures between 1,085 and 2,174 degrees F. Lead-containing glazes can appear to mature at these temperatures, but firing temperatures that assure safety vary greatly, depending on how the glaze is formulated. For example, a lead glaze formulated for an optimum firing temperature of 1,975 degrees F will be underfired at 1,830 degrees and will not be stable enough to keep lead from leaching into food and liquid. Potters in foreign countries often use wood-fired kilns, in which temperatures are difficult to regulate. Even gas or electric kilns are sometimes stacked with ware beyond their capacity, and small items may be placed inside larger ones, making it impossible for each piece to receive consistently even heat.

Robert Harrison, assistant head of the ceramics program at the Banff Center School of Fine Arts in Alberta, says that despite their higher cost and drawbacks when it comes to producing bright colors, nonlead glazes are now being used by most, if not all, independent potters, and comments,

> I would say that the studio potters here in Canada use almost exclusively nonlead glazes for functional food surfaces…There has been a growing awareness of the dangers of lead among the general public over the 15 years that I have been involved in ceramics, and there isn't a single studio potter I know of who uses lead glazes for functional food surfaces.[49]

Though independent craftspeople seem willing to sacrifice color and economy by using nonlead glazes, almost all foreign pottery, which is often bright in color, and that produced by domestic manufacturers is still coated with lead-containing glazes. "No one has yet discovered a substitute material that will give you the brightness, the depth, everything that a lead glaze will give you," says Robert J. Beals, director of research and development for the Hall China company. "If you are going to operate at a low temperature, like 2,000 or 2,100 degrees F, then the only material that will give the bright glaze needed is a lead-containing glaze."[49]

A. How Much Is Too Much?

Approximately 1,000 tons of lead are mined in Canada every day. Most of it finds a safe and useful home in numerous products—about 60 percent of it is used in batteries—and remains more or less where it is put. But some does not. Lead in gasoline, for instance, is gradually being phased out because when the gas is burned, lead is released into the environment. In the 1970s, lead levels in paints were restricted in Canada after researchers found that paints on toys and cribs were contributing to lead poisoning in children. Even now, however, the regulations are not foolproof. The U.S. Consumer Product Safety Commission recently recalled 100,000 toy brooms marketed by Come

Play Products of Worcester, Massachusetts, because they were painted with lead paint.

Despite new regulations on lead in the environment, the metal is still finding its way into our bodies, raising the total body burden of lead for some persons to levels that, according to several scientists, can lead to chronic kidney disease, hypertension and, in severe cases, death from lead intoxication itself. In general, the levels of lead in the blood of people in North America have been dropping. This is mainly due to the reduction of lead in gasoline. The problem, however, arises with unexpected exposures that are occurring and are not detected. These exposures can arise from several sources, including pottery, contaminated water and lead paints, and are not at all rare today.

In those instances where ceramics are involved, the exposures are much higher than the levels found in the general population of 20 years ago and are two to three orders of magnitude higher than the background levels today. In fact, they are comparable to occupational exposures known to cause chronic conditions such as kidney disease and hypertension.

Five percent of the 80,000 Europeans currently being treated with kidney dialysis machines are diseased due to lead exposure, and future studies may show the same to be true in North America. Kidney dialysis in the United States costs $30,000 per person per year. This means that the cost to society of treating the 4,000 U.S. citizens who are diseased by lead is $120 million per year.

There are several avenues through which lead is still entering the food chain and the water supply. Lead-containing pesticide residues in fruits and vegetables, lead water piping that is still in use in some old buildings, lead wall paint that has been painted over rather than removed, lead in the runoff from mining, industrial dumping and in food packaging (mainly lead solder in tin cans) all contribute to the background level of lead and, consequently, to our bodily intake.

The difficulty of documenting the actual amount of lead that resides in a human body and the lack of knowledge concerning the physical consequences of even small amounts absorbed over many years has led to the establishment of what are called "acceptable" or "normal" levels. These levels are measured by "blood-lead" tests and have been the basis for deciding how much lead a person may consume each day and still be "safe." For example, a study of 1,315 children in southwestern Ontario by the Ontario Ministry of the Environment, showed by 65.3 percent of children under six years old in the Toronto-Windsor corridor had blood-lead levels in the 10 to 19 micro-grams per deciliter range, and 5.6 percent of urban children over six had blood-lead levels over 20 micrograms per deciliter, the "alert level" determined by the Toronto Board of Health.[49] "If these percentages are applied to all children of this age in Canada," say Barbara Wallace and Kathy Cooper in their book *The Citizen's Guide to Lead*, "it can be estimated that about 35,000 children have 'elevated' blood-lead levels, over 100,000 children may be above the 'intervention level,' and almost 1.3 million may be at risk of minor developmental deficiencies due to chemical changes." The classic symptoms of mild lead poisoning include hyperactivity, behavioral changes, inability of the blood to create hemoglobin and a lowered I.Q. Admittedly, 60 percent of our blood-lead comes from gasoline, but an increase in lead in the diet could be enough to put many children over the limit.

Because lead moves out of the blood fairly rapidly after being consumed, however, and accumulates and remains in the bones and organs for at least 30 years, many researchers are now convinced that blood-lead screening is a totally inadequate method of determining how much lead is actually in the body. There are many people with a deep vested interest in the blood-lead test. These include physicians who do not want to admit that they have been wrong about this for 50 years and those whose interests are tied up with the lead industry itself. But the fact is that the test fails in those areas where lead exposure occurred a long time ago or over long periods of time.

Lead leaching from the glaze of pottery is a serious public health hazard, one that not only poses the threat of lead poisoning, but also contributes significantly to the overall body burden of the metal for

THE LEGACY OF HEAVY METAL POISONING

many persons, a burden that is being under-estimated by government and many physicians due to their reliance on blood-lead tests.

The deleterious effects of lead in the diet have been known for centuries. Gout, common in Europe for centuries and originally thought to be caused by overeating, has been linked to lead in the diet. Some researchers are now convinced that dietary lead contributed significantly to the fall of the Roman Empire, causing sterility and mental disorders among the aristocracy who drank wine stored in lead-lined vessels and water conducted through lead pipes. Food for the aristocracy was cooked and served almost exclusively in leaded cookware, and some dishes were even sweetened with a lead-laced concoction called *sapa.*

Among these researchers is Jerome O. Nriagu, a research scientist with Environment Canada and author of the book, *Lead in Antiquity.* "During this period, most eating, cooking and drinking utensils were made of lead pewter," he says. "And gout, which is one of the clinical symptoms of lead poisoning, was extremely widespread and common." Nriagu is also concerned about the amount of lead in the environment today. "What should happen to us if we keep contaminating our environment with lead?" he asks. "Shall we go the way of the Romans? I take the view that no lead is good lead."[1]

Despite all the historical evidence concerning the adverse effects of lead on human physiology, efforts to remove the metal from the environment have been only partially successful. And most of us are still unaware of the danger. There is more information on the dangers of lead than on 90 percent of the environmental hazards that hit the headlines every week, and though the negative evidence concerning lead is probably much stronger than that for all the organic compounds the public is frightened of, such as formaldehyde, dioxins and many pesticides, lead still remains a boring subject—the issue does not have as much "sex appeal" as something like AIDS.

One of the major difficulties in dealing with severe lead intoxication is that the symptoms can mimic those of many other diseases and health problems. The discovery that a person is suffering from lead poisoning is often made only after considering numerous other possibilities, and as a result, many people have died or come close to death before a correct diagnosis was made. Though many of these cases have occurred in those who are employed in lead or lead-related industries, there have been several instances of lead poisoning from dinnerware reported in the medical literature. For the most part, the sources of lead are not well recognized. Doctors do not suspect that the cause of the illness might be lead intoxication.

Don Wallace and his wife Fran know the consequences of such misdiagnosis firsthand. Before settling in Seattle, Washington, and opening his laboratory, Wallace was a career officer with the U.S. Air Force. In 1977, the Wallaces were stationed in Italy, where they bought a set of locally made dishes. Shortly thereafter, Don began experiencing severe pain in his arms, diarrhea and abdominal cramps. Doctors examined him but found no reason for the strange symptoms. Though they used the Italian dishes only occasionally, both Don and his wife daily drank coffee from the cups in the set and soon Fran, too, became ill. Both began to lose weight and to have trouble sleeping, and Don's personality began to change—he became more and more irritable and aggressive. Before a satisfactory diagnosis could be made, however, the Wallaces were sent back to the United States. The dishes were packed away, and they started feeling better.

While in the States, they did some traveling and used the Italian pottery only sporadically. After settling for a time in Monterey, however, Fran became severely ill and was rushed to the hospital, suffering from terrible pain, weakness in her legs and an inability to eat or keep down liquids, causing her to become dehydrated. Doctors told the couple that Fran was suffering from acute intermittent porphyria, an inherited enzyme deficiency in the blood for which there is no cure. While in the hospital, however, she began to feel better and was eventually released.

Following her hospital stay, Fran was weak but no longer in pain. Next they were sent back to Washington, D.C., and then on to the Dominican Republic, where they started using the dishes again and began to experience the same set of odd symptoms. This time, however, they chalked it up to their lack of experience in a tropical setting. "We just kept saying it was the heat," says Fran, "that's why we couldn't eat and were feeling weakened." The Wallaces endured the situation as long as they could, but finally, Don took an early retirement and they moved to Seattle, where, after about six months, they were actually feeling pretty well. But then the nightmare started all over again: within three months, Fran was back in the hospital with what appeared to be a fatal attack of porphyria.

"The physicians didn't know very much about porphyria," she says. "I was vomiting, and they could do nothing for the pain, which was horrendous. About all they could do was try to keep me from dehydrating. Also, I had developed a strange anemia, which they seemed unable to explain." By this time, Don had begun to lose faith in the doctors' ability to cope with his wife's problem, and he started his own investigation into the possible cause. "He started going to the university every day," Fran recalls, "and combing the literature in the medical library for clues." Knowing that the one anomaly in his wife's diagnosis was the anemia, Don concentrated on anemias and their causes in addition to porphyria. But by the time he found what he was looking for, he was also severely ill. "He was a skeletal weakling," Fran comments, "unable to sleep or eat, and he had been told that I was dying. Then he came across a paper mentioning that lead poisoning can mask itself with the same symptoms as some anemias. He brought the paper to the doctors, but they just said that the idea was ridiculous and refused to even test me for blood lead. It seemed to me that they had taken my blood out by the barrel daily, but even with all the barbaric tests they put me through, they had never done a heavy metals test."[49]

Finally, Don enlisted the help of a neurologist and geneticist at the university. Together, they confronted Fran's team of doctors and succeeded in getting the tests run. The diagnosis immediately changed from porphyria to lead intoxication. Don was then tested, too, and found to be dying from lead poisoning. Only their son, who did not drink coffee, tested normal.

Therapy to remove lead from their blood restored the couple to good health, at least temporarily, and Don turned his investigatory skills toward finding the origin of the lead that had nearly killed them. Noting that the symptoms seemed to have disappeared when Fran was in the hospital and when their belongings were in storage, he decided to have tests run on the Italian dishes. Sure enough, tests showed that the dishes were leaching sufficient lead into their food and coffee to poison them. And because their symptoms were misdiagnosed for three years, this poisoning had gone untreated until they were both near death.

Though they have been free of symptoms since the incident, Don and Fran are both aware of the long-term consequences of lead in the body. "Researchers now know that lead is stored in the organs and bone marrow, and that it can be very damaging to the kidneys," says Fran. "We may still be looking at some future problems. We just don't know what it might have done already or will do later."

Since recovering from his ordeal, Wallace has gone on to get a master's degree of science in public health from the University of Washington and now dedicates most of his time to alerting the public to the threat of lead in pottery glazes (it was his laboratory work that sparked the Williams-Sonoma and Pier One recalls). Some lead-glazed pottery, he believes, can be found in nearly every home in North America.

B. The Controversy

Representatives of the pottery industry are concerned that the release of information about the dangers posed by lead in glazes could frighten the public regarding the safety of domestic dinnerware. Researchers with ties to the lead and pottery industries have continually disputed published scientific reports that suggest the allowable limits are too high. The main disagreement concerns whether a vessel releasing lead in amounts now allowed will deteriorate with use and begin to leach more than it did when tested.

Robert Beals, who in addition to his duties at Hall China serves as research chairman for the U.S. Potters' Association, believes that if a vessel is tested and found to be within the acceptable limits, it will remain so, even if scrubbed with scouring pads and washed in the harsh detergents used in dishwashers. Beals explains,

> We have run tests on domestically produced ware bought at commercial locations...We have scoured them with steel wool and other agents to give them as rough a test as could possibly be done short of grinding them up into fine powder. Even after this harsh treatment, we were not able to discern any increase in lead release from properly fired, domestically produced work.[49]

Don and Fran Wallace have tested hundreds of dinnerware and tableware items in an effort not only to locate imports that evade government screening programs, but also to compile information on the amounts of lead that will leach from pottery under various conditions. According to Wallace, if dinnerware is found to leach far below the CCAC limits (below 1 ppm), as does most of the ware manufactured in North America, it is unlikely that the glaze will break down with use. However, items that leach just below the maximum now allowed can deteriorate with use and begin to release several times what is legally allowed.

> We have tested a number of Italian bowls that, in new condition, released amounts of lead within the limits set by the FDA for admission into the U.S. After scrubbing these bowls the way you would if you were trying to remove stains, however, the Lead release increased by two to three times, causing them to drop below FDA and CCAC standards.[49]

47

In 1984, an in-house paper issued by the U.S. Centers for Disease Control in Atlanta, Georgia, stated that

> ...repeated washing has been shown to promote deterioration of glaze such that pottery previously tested as safe can become unsafe for use... Consumers should not use pottery for the cooking or storing of food or beverages unless recently certified free of leachable lead.[49]

With hundreds of millions of foreign pottery items being imported into North America annually, even if only 1 percent leach close to but below the limits, it would mean that millions of potentially dangerous items are being sold to consumers each year.

Wallace believes that lowering the lead-release limits would reduce the number of pieces that fall into this "grey area" and would eliminate the contradiction between the goals now set for lead consumption and the amount of lead allowed to leach from pottery. Domestic manufacturers, he says, should have no problem with lowered limits as they consistently produce ware that leaches far below the current limits. Most importantly, tougher standards would help eliminate the number of foreign items that might become dangerous through use. Robert Harrison at the Banff Centre agrees that the limits should be lowered. "With all the technology and the engineering and chemistry information we now have," he says, "there ought to be no problem in lowering these limits or in eliminating lead from food surfaces altogether."

Though present CCAC standards say that glazes used for food surfaces can contain lead if they do not leach in excess of certain limits, a paper issued by the FDA on March 18, 1985, appears to contradict this. The paper, which recounts the Wallaces' experience, states that "pottery with lead glazes is suitable for ornamental purposes but poses a hazard when it is used for food preparation, storage or service." And it concludes by advising consumers "to examine their collections of mugs and dinnerware and any handmade or unusual pottery or other ceramic containers and to use for foods only those items they can be sure are free of lead."

Industry, government and public health advocates are all concerned about pottery that is carried into Canada by tourists who may have bought it from an ethnic potter using improper glaze formulations and firing techniques. Another area of concern to all is the individual ceramist who may be improperly formulating and firing lead glazes for functional food surfaces. Finally, all agree that antiqueware and some ware manufactured prior to 1970 can be hazardous.

According to Wallace, tests to determine how much lead is likely to leach from a piece of dinnerware are simple and can be performed at almost any university or private chemical laboratory. If you have an item you are concerned about, he says, the best thing to do is either have it tested or use it only for decorative purposes, not for food preparation, serving or storage.

Among the most dangerous of all pottery is that which is brought in from Mexico. Several pieces of Mexican pottery have been tested by Wallace as well as by CCAC and have been found to release hundreds of times the maximum allowed. Additionally, surveys conducted in Mexican potters' villages have shown consistently high levels of blood lead in the inhabitants. Even those villagers who did not work directly with lead glazes were found to have high levels of lead in their blood. Ed Steel, Director of the Division of Program Operations at the U.S. Center for Food Safety and Applied Nutrition, comments

> The People's Republic of China had a tremendously high violation rate, so high that we invoked an automatic detention on the entire country. What this means is that we automatically deny any shipments from China until they have been tested by an independent lab that is suitable to us.

Murray Kaiserman of CCAC is aware of the problem with Chinese pottery. "Our two biggest violators right now," he says, "are Mexico and China." Wallace sees the problem of lead leaching from glazes as analogous to a slow-motion shotgun, aimed at the public and going off one lead pellet at a time, commenting

> With all we now know about the short- and long-term dangers of lead in the body, it seems to me that the most ridiculous place to put this highly toxic and cumulative substance is on the surfaces of food containers, cooking utensils and dinnerware used regularly by the consumer. To my mind, lead in the glaze of pottery has been and still is a major threat to public health.[49]

Independent Enquiry on Lead Danger by CBS News, April 5, 1988

Congress was supposed to get a new report this week on the effects of lead on children, but it won't. Its authors say the report has been buried under bureaucratic red tape. Consumer correspondent Aaron Moriarty, though, managed to get a look at the report and some of its frightening conclusions.

An estimated 3 to 4 million American children have damaging levels of lead in their blood. That statistic may come as a surprise to people who think that the lead problem is behind us, but this latest report on lead says the government hasn't done enough, that there's much more lead left in the environment and more damage from the lead than most of us realize. Last year more than 20,000 pieces of ceramics and glassware had to be pulled off store shelves. The problem? An excessive amount of lead in the paint and glaze. Lead is also in drinking water, in as many as 20 percent of our homes and in schools. In Los Angeles, where officials are currently testing school water fountains, anywhere from one-fourth to one-half of those tested dispense lead contaminated water. It is in food, imported canned food.

In a test conducted by CBS News on products bought in New York stores, 13 out of 16 samples contained detectable levels of lead. Add to that an estimated 27 million homes and apartments that still have lead paint, communities with lead smelters, and air filled with leaded gas emissions and you have what the soon-to-be-released government report says is a far greater health threat than the public realizes.

We're talking about 3 to 4 million young children, okay, who have blood lead exposures that are unacceptable. Whether they're urban, suburban, well to do, rich or middle income or poor, it affects them all. The report, already delayed more than a year, was supposed to be given to Congress, but officials at Health and Human Services have refused to release it, or even schedule a date. Government officials say that they need more time to analyze the information in this report, but those who prepared it say that information is simply too embarrassing.

The abatement efforts that have been done by various federal agencies and Congress and local and state agencies have been mixed, at best, and abysmal at worst. One example is the FDA Food Evaluation Program. While the amounts of imported foods coming into this country increased 14 percent last year, the FDA does not test these foods for lead content, even though a large portion comes in cans sealed with lead solder. They do not regulate, or even examine, the imported canned foods. Tests show that over 80 percent of samples contained lead, 12 percent with levels that exceed EPA standards.

Some of these values, I believe, are very high. For instance, the hearts of palm (a canned root of palm) are enormous; it's 14 parts per million. That is so high that it is probably unacceptable. We've been forced to cut back our import surveillance. That's just a matter of the resources that are available to the agency. The government only spot checks imported items, even though the FDA's own research conducted last year indicates that nearly 10 percent of imported ceramics may release lead into food.

While the lead from imported pottery in foods rarely causes severe lead poisoning alone, when that amount is added to the lead you get from air, from drinking water, and even dirt, then it can become a health hazard, especially to children. Problems in learning, problems in school, problems in emotional and behavioral development, problems, in fact, even in rate of growth may all be expressions of lead. And these problems are caused by levels of lead below what was once believed to be dangerous.

We've been telling you about a brand new government report on the dangers of lead, a government report that the government refuses to release, but Aaron Moriarty got a copy of it and she has more on the dangers of lead you can encounter in your own home. If you use dinnerware mugs that have been in the family for years, or you drink your coffee and orange juice in imported pottery, you could also be getting lead in your diet. Now, no amount is safe; too much is down right dangerous.

CBS Interview with Two Lead Poisoning Victims

I started feeling a little worse and a little worse and I finally went to the doctor who said I had a virus and, you know, it was all the symptoms of a virus. I was run down and just felt rotten. I went to eye, ear, nose and throat doctors and they tipped me upside down and ran all kinds of tests and now they couldn't understand what was wrong.

What doctors thought was just a cold was, in fact, lead poisoning. Washington State residents Jim and Pat Apperson were slowly poisoned every time they used 40 year old drinking glasses they inherited from Pat's mom.

How often did you use these glasses?

(Mrs. Apperson) Well, every morning for milk, and probably tomato juice or grapefruit juice. Basically, they were used constantly, we used them for our everyday glassware in the kitchen, for drinking glasses.

Such severe lead poisoning is rare, but pottery and glassware containing lead are not. Lead paint and glaze were commonly used on items made in this country before 1970 and are still used on imported ceramics. When those pieces are improperly made, fired at temperatures below 1200 degrees C, the lead can be released into food. The most common sources of substandard pottery are Mexico and Italy. All of these have been recalled from the retail shelves of Pier 1 Imports in the last couple of years. The blue pitchers, the blue bowls, and particularly some of the yellow pitchers were found to release very high levels of lead.

The federal government also checks imports for lead, but only spot checks them, and critics say potentially dangerous products slip by such as these drinking glasses given away last summer as part of a service station promotion.

There are over 300,000 shipments of pottery coming into the United States per year. A shipment could be as large as a cargo ship. With only 850 inspectors, it's impossible to check everything coming into the United States. Lack of government testing has forced American companies like Pier 1 to begin testing imported products on their own, but the Appersons, who have experienced the effects of lead poisoning, do not feel safe. It has taken two years of blood transfusions and the removal of Jim's gall bladder to give them back their health.

Lead in Tap Water, Ground Soil and Paint

Maurice Sanders and his wife, Judy Southerland, were worried. Their ten month old daughter Olivia was having problems. She had virtually stopped growing, dropping from the thirtieth to the fifth percentile of her age group in height and weight; she suffered from constipation, insomnia, and crankiness. At first, doctors couldn't account for her condition. Then a routine blood test in October 1985 yielded a surprising result: Olivia was suffering from lead poisoning.[50]

Her parents were shocked. Wasn't lead poisoning an affliction striking inner-city children who chew on peeling lead paint? How had their child, growing up in an affluent part of Washington, D.C., become contaminated? Sanders and Southerland spent $4,500 to strip and repaint their house and replace the drywall ceilings, just in case any lead-based paint had ever been used. They tossed out newspapers and magazines because lead pigments are used on color pages, which a baby might nibble. But Olivia's high lead count persisted.

Nearly a year after their daughter's first blood test, Sanders and Southerland decided to have their tap water checked. Several separate samples showed lead levels nearly four times higher than the 50 parts per billion the Environmental Protection Agency (EPA) considers safe.

The family had suddenly become part of a silent epidemic of alarming proportions.[50] According to a growing body of medical research, millions of Americans, most of them children, are suffering from low-level lead poisoning. New studies have uncovered evidence of serious physical and intellectual impairment in children with only small amounts of lead in their blood—amounts far below the 24 micrograms per deciliter officially considered the maximum safe level by the Centers for Disease Control (CDC). In fact, a March 1986 report from the EPA's Clean Air Scientific Advisory Committee suggests lowering the maximum safe blood lead level from 24 micrograms per deciliter to 9. This widespread contamination springs, not just from crumbling paint, aging factories, or common tap water, but from such sources as the food we eat and the dishware we use, the dust in our homes and the dirt in our yards.

The body doesn't need lead, yet everyone absorbs it in varying amounts; some of us tolerate it better than others. Lead isn't excreted, but is stored for many years in tissue, chiefly in the bone, from which it is released back into the blood-stream to wreak cellular havoc. Especially vulnerable to lead overload, according to the EPA report, are millions of Americans in three groups. For middle-aged men, the risk of hypertension rises along with blood lead levels. According to a 1985 study by the CDC, levels as low as 5 micrograms per deciliter can increase blood pressure. A second group at risk is pregnant women, for whom lead means a greater chance of miscarriage, premature delivery, and stillbirth.

Perhaps most disturbing is the danger to children between the ages of five months and six years. According to a report about to be submitted to Congress by the Agency for Toxic Substances and Disease Registry of the Public Health Survey, an estimated 17 percent of preschool children have blood lead levels exceeding 15 micrograms per deciliter. At that level, children are susceptible to a range of psychological, neurological, kidney, and blood abnormalities, including partial hearing loss; slower neural transmission; hyperactivity; learning disability; lower IQ scores, impaired ability to metabolize vitamin D, absorb iron, and use calcium in any bodily processes; disturbances in the formulation and maintenance of red blood cells; decreased muscle tone; and interference with the creation and function of certain enzymes and amino acids. New studies show that high maternal lead levels can damage a child prenatally, lowering birth weight and causing later deficits in physical and intellectual development.

Historically, the federal government has avoided regulating a metal of such commercial and military importance, although its toxic effects have been known for millennia. (Some historians theorize that lead led Rome to ruin. Poisoning from extensive use of the metal in utensils, weapons, cosmetics, wine vessels, and water pipes may have been responsible for imperial madness, infertility and miscarriage rates that kept the ruling classes from replacing themselves).

The leading of America began in earnest with the Industrial Revolution. Smelters and manufacturing plants spewed fumes with abandon, and workers toiled unprotected from the mid-1800s until early in this century, when occupational regulations were instituted.

In the 1920s, who knew that adding lead to gasoline would backfire? Adding it to paint for longer-lasting freshness seemed like another great idea, and lead solder was a good sealant for canned foods. It took the social activism of the 1960s and early 1970s to bring about controls; finally, in the mid-1970s, the Food and Drug Administration (FDA) pressured the food industry to remove lead solder from the seams of baby food and baby formula cans;[23] and, the Lead-based Paint Poisoning Prevention Act banned the manufacture and sale of leaded paint, directing the Department of Housing and Urban Development (HUD) to develop a strategy for removing old paint from old housing.[108]

According to the EPA, more than 161,000 metric tons of lead tainted the air in 1975. In an effort to lower this atmospheric pollution, the agency ordered oil companies to start phasing out lead additives in gasoline beginning in 1977;[113] by 1984 the count had fallen by 75 percent to 39,000 metric tons. Each year since, the levels have continued to fall.

The effect of these reductions on blood lead levels was dramatic. In 1982 the National Health and Nutrition Examination Survey recorded a 37 percent drop in average U.S. blood lead levels between 1976 and 1980. But even more dramatic was the discovery of how big a problem still remained. The survey showed that, even with the drop, unexpectedly high levels of lead were coursing through the veins of an unexpectedly high percentage of the population. About half the adult U.S. population surveyed had blood levels above 10 micrograms per deciliter. Eighty-eight percent of the preschool children tested had blood levels at or above that level, and 9.1 percent of them met the current CDC criteria for lead poisoning. The numbers were even more alarming for black children: nearly a quarter were lead poisoned, and 97.5 percent reached or exceeded 10 micrograms per deciliter.

In 1985 gasoline still spewed close to 20,000 metric tons of lead into the air, enough to cause an estimated 123,000 cases of high blood pressure a year. In 1987 lead paint still covers 25 million housing units. Solder in cans still accounts for up to 24 percent of low-level lead poisoning. The amount of lead in ash generated by the nation's 73 resource recovery incinerators—ash often hauled to disposal facilities in open trucks—will soon dwarf the amount of lead used in gasoline, even as lead levels in gas fall. The popularity of incineration as a method of waste disposal continues to rise. And in July of this year, the Occupational Safety and Health Administration cited and fined the Chrysler Corporation for 811 job-safety infractions, including willfully exposing autoworkers to lead and arsenic.[50]

The EPA decided to survey the nation's tap water; it reported surprisingly high levels. Tap water, it found, represents 15 to 40 percent of the lead to which Americans are exposed, depending on local conditions. The problem seems to lie, not in municipal treatment plants or public distribution lines, but in residential plumbing systems, where water collects small particles corroded from lead pipes or, more often, lead solder in copper pipes. More lead enters the taps of communities with soft water (low in magnesium and calcium) or acidic water, both of which are corrosive, and more is found in new homes, because freshly applied solder dissolves easily.

Another newly considered source of low-level lead poisoning is plain old dirt. Auto emissions, flaking paint, and smelters left a legacy of 4 million to 5 million metric tons of lead in dust, soil, and sediment. Children encounter it in backyards, fields, and playgrounds, especially those located near highways or smelting plants.

When Denise Wadleigh of Jersey City, New Jersey, was pregnant with her fourth child, she and her husband bought a home in a middle-class residential neighborhood not far from a major highway. Less than a year later, a routine blood test showed that Wadleigh's two year old son, though asymptomatic, was suffering from lead poisoning. A scientist from Rutgers University analyzed the backyard dirt and found lead levels as high as 1,800 parts per million-100 times higher than average world soil levels. "We had to spend six thousand dollars to cover the backyard with concrete, for the safety of my children," Wadleigh reports.[50]

Elsewhere the problem of lead-poisoned soil has become so severe that whole towns are affected. In October health officials uncovered 28 cases of lead poisoning among children living in a low-income section of Thompson, Connecticut.[50] In all of Thompson in the previous 18 months, a staggering 21 percent of children tested positive for lead poisoning. Investigators have traced the source of the problem to Thompson's soil, but no one has yet determined how the metal got there.

Powder-fine outdoor dirt sifts inside to become household dust, presenting a danger to children who crawl on the floor and gnaw on furniture. Such dust can also contaminate fresh vegetables and fruits, especially those grown near highways, though a thorough washing ought to remove about half the lead.

Other, less widespread contributors to the problem include lead cooking utensils and lead-containing fertilizers and pesticides. Even dinner plates, cups, and glasses pose a danger. Molecules of toxic lead can dissolve from the glaze of imported ceramic ware that hasn't been fired at the high temperatures legally required in the United States.

Don and Fran Wallace of Seattle learned that the hard way. In 1978, when Don was an Air Force lieutenant colonel stationed in Italy, the couple bought a charming set of terra-cotta dishes in a small village. Soon after the purchase, Don became uncharacteristically irritable and aggressive. He lost more than 30 pounds and suffered from insomnia and pains in his wrists and forearms. By 1981 the Wallaces had moved to Seattle, where Fran became gravely ill with body aches, anemia, and dehydration. After searching through medical books, Don insisted they be tested for lead poisoning. The two were found to be severely poisoned. By the process of elimination, Wallace traced the source to the Italian dishes, especially two coffee mugs that, tests showed, released over 300 times more lead than FDA standards permit.[23]

Imports account for more than 60 percent of U.S. ceramic dinnerware sales, and the FDA has stepped up inspection of foreign-made dishes, stopping more than 1,000 shipments last year after they were discovered to release lead. Most violations were found in products from developing countries.[4]

How much lead is harmful? Until 1970, contamination was considered dangerous only at blood levels associated with extreme symptoms of poisoning: convulsions, brain swelling, and acute kidney disease in children who registered 80 micrograms or more of lead per deciliter of blood, stomach pains, hallucinations, and wrist weakness in factory workers with the same elevated blood lead levels. But as Olivia Sanders' case[50] shows, low-level lead poisoning can cause symptoms that, while more subtle, are no less serious.

When doctors began identifying symptoms at significantly lower blood levels in the 1970s, the Centers for Disease Control kept lowering the point at which patients should be treated for lead poisoning. Today many physicians consider the current CDC standard of 25 micrograms per deciliter to be medically obsolete.

The most influential early study of lead's low-level effects was published in 1979 by Herbert Needleman, then at Harvard, now professor of psychiatry at the University of Pittsburgh Medical School. Testing first and second grade students in the Boston area, he found the lowest IQs, academic achievement, language skills, and attention span among children with the highest body burdens of lead. None of their lead levels exceeded the CDC's toxicity threshold. "Until then it was generally thought that a child had to be sick to lose intellectual function or have behavior disorders," says Needleman.[50]

Instead of testing blood samples, which was the norm at the time, Needleman measured lead levels found in the children's baby teeth. He found that children with dental lead levels of 10 parts per million or less had IQs averaging four points higher than those with levels of 20 parts per million or more. Worse, he found that the children with higher lead levels were nearly four times more likely to have IQ scored below 80 and seven times more likely to suffer learning disabilities. No children with lead levels between 15 and 35 parts per million scored in the superior IQ range of 125 or above, while 5 percent of children who were virtually unexposed to lead did score this high. "If that 5 percent were to be lost, too," Needleman says, "it would be an enormous social loss."[50]

The small amount of research done so far on lead's long-term effects suggests that intellectual impairment may be irreversible.

When the first graders in Needle-man's study were retested five years later, those who had originally shown higher lead levels still exhibited significant IQ deficits and required more special education classes than did the children who had been considered virtually unexposed. The latest evidence suggests that lead can imperil unborn children who absorb the metal from mothers exposed to even small doses. A study reported in the New England Journal of Medicine by Needleman and others[50] found that children who had absorbed the most lead while in the womb performed significantly worse on developmental tests in their first two years of life than did children with low exposures.

Not surprisingly, the lead industry refutes new findings about the sources and effects of low-level lead poisoning. Werner Meyer, Lead Industries Association president, notes that although lead causes measurable physical changes, there is no proof those changes are harmful. He bristles at "nonscientific accusations by extremist environmentalists intent on grabbing headlines," and claims "the amount of lead in water, the latest hue and cry by pseudoscientists, is vastly overblown." Most lead in the environment, he adds, comes not from man, but from eons of volcanic eruptions.

The global record of lead emissions, as revealed in polar snow strata, ocean sediments, tree rings, and skeletal remains, tells a different story, however. It shows two big leaps in atmospheric lead levels—one at the start of the Industrial Revolution, another in the 1920s, when lead was first added to gasoline.

In some cases the treatment for low-level lead poisoning can be as simple as removing the victim from the source of pollution. For Olivia Sanders, that was enough. Her symptoms began to abate two months after she stopped drinking tap water; she began eating and sleeping better, became less irritable, and stopped losing weight. After six months she'd gained height and weight and become more outgoing. No one knows what the long-term effects of lead poisoning may be for Olivia or her twin. Nor does anyone know how to ensure that lead victims like Olivia have been entirely removed from the source of contamination. Indeed, it is often impossible to predict just what water will be safe.[50]

Jeanne Briskin of the EPA reports that in uncounted schools across the country, the tanks used to store drinking water for fountains may be entirely lead lined. Even the child whose parents have removed the lead threat at home may still be taking great gulps of the metal dozens of times every school day. Children with high blood lead levels—usually above 35 micrograms per deciliter—and acutely poisoned adults are treated with chelating drugs, chemicals that speed the excretion of lead from the system but have dangerous side effects. Denise Wadleigh's son required chelation, as did the Wallaces.

Early detection is vital, say the American Academy of Pediatrics, which recommends regular screening for pregnant women and young children; men at risk for hypertension should also have their blood levels tested regularly. Prevention is even more important. For the individual, that means testing potential household sources of pollution and eating right: a diet with adequate protein, low in saturated fats, and rich in minerals like iron, zinc, and calcium helps protect against lead poisoning.

Is it possible to lead lead-free lives? "If we cleaned up the environment and optimized nutrition," says Needleman, "blood levels could drop as low as three to five micrograms per deciliter—what they are in primitive cultures."[50] Public health specialists have intensified efforts to control the pollutant. The EPA plans to reduce by 60 to 80 percent the amount of lead permitted in drinking water, from 50 parts per billion to no more than 20. The agency ruled that most water companies must inform customers of lead levels in their drinking water and of the dangers posed by the metal. The leaded gas phase-out is supposed to continue, and the agency's Superfund is to provide $15 million for a demonstration project to test the feasibility of soil cleanup. The FDA is pressing for further reductions in lead solder in cans, and HUD has finally come up with plans to survey lead paint in federally insured, owned, and subsidized housing.

Some physicians are surprised that this silent epidemic hasn't caused more of a stir. "I think people still see it as a disease of minorities," says Needleman. "But we've known about the gross effects of lead for centuries." Paul Mushak, adjunct professor of environmental pathology at the University of North Carolina School of Medicine and a principal co-author of the Agency for Toxic Substances report, agrees with Needleman and confesses impatience with people who quibble over the magnitude of the problem. "Is it tolerable that our kids aren't as bright as they could be?" he asks. "If it doesn't put people in the hospital, it isn't a serious threat. If you had children with scarlet fever or whooping cough at these numbers, you'd have a mutiny in the streets. Our children and people in risk groups shouldn't be held hostage to a preventable disease."[50]

Most experts see government intervention as a solution to the lead problem: regular blood screening for children and pregnant women at high risk, injection of chemicals into public drinking water to reduce corrosiveness, elimination of lead additives from gasoline, and removal of polluted topsoil from playgrounds and school yards. Each of these steps, however, requires that the government take direct action, something it generally does slowly at best. As usual, therefore, the first and fastest line of defense is consumer awareness.

Here's what you can do:

* Check for chipping, flaking, or cracking paint in your home. Have the chips tested by your local county health department. If the chips contain lead, remove and replace the paint immediately. If you remove the paint yourself, use a sander and wear a surgical type mask. Don't use a heat gun, which torches the lead into toxic fumes.

* Since lead pollution of water occurs most often in residential plumbing and in service lines, the pipes that link residences to the municipal water supply, it's important to test water directly from the tap. It is

important to test water directly from the tap (from a first draw water that's been sitting in the pipes all night and again during the day). Homeowners unable to get their city or county water departments to do the testing can pay to have a private laboratory test the water. If there is lead in the line, a stopgap remedy is to run the tap water for three minutes before using it, to flush out accumulated contaminants. This must be done for every drinking faucet. Be sure to use water only from the cold water tap for drinking, cooking, and especially for making baby formula. Hot water is likely to contain more dissolved lead. The installation of carbon filters, sand filters, or cartridge filters is generally not helpful since these devices remove some water contaminants but not lead. More effective cures include re-placement of lead joints and pipes with ones made of copper (be sure to use lead-free solder), and the use of bottled water.

* Overseas travelers, especially in Third World countries, should beware of ceramic dinnerware; its lead glaze may not be fired at high temperatures required to prevent chipping, flaking, and leaching of lead particles into food. The same caution should be extended to dishes manufactured in the United States before the 1970 government crackdown on poorly fired domestic and imported dishware.

Uranium

A. Fiesta Ware

Fiesta was a very popular art deco line of dinnerware that, in the 1930s and 1940s, was estimated to grace the tables of nearly one-third of American households. Produced by the Homer Laughlin China Company of West Virginia, Fiesta was avail-able in several extremely bright, highly glazed colors. The chief component of its most popular shade-mango red—before 1945 and from 1959 to 1972, was uranium oxide.

Known to collectors as "Radioactive Red," a set of Fiesta dinnerware could actually set off a Geiger counter; high school science teachers used it as a convenient source of radiation for classroom experiments. Since the 1960s, Fiesta has been the subject of repeated investigations by the United States Food and Drug Administration (FDA) as well as the Nuclear Regulatory Commission (NRC), and as recently as 1981, the New York State Health Department recommended that consumers keep all uranium-glazed orange-red ceramic tableware—other producers of the pre-War era included Caliente, Harlequin, Poppytrail and Franciscanware—off the table and away from food.

Homer Laughlin, the world's largest manufacturer of dinnerware, estimates that until 1943, it was using about 90 percent of the uranium being produced in the United States in its glazes. A company history tells of an incident in which federal agents, looking for sources of uranium for the atom bomb, appeared at the company's West Virginia plant, incredulous that so much uranium was being used to manufacture dinnerware. Without divulging the reason, the agents confiscated Homer Laughlin's uranium supply and told the company that the element would not be available to them in the near future. Unable to obtain the key ingredient for their glaze, Homer Laughlin had to discontinue mango-red Fiesta. Production was resumed in 1959, however, when uranium became commercially available again. The amount of uranium in the glaze was reduced only when NRC regulations limited its quantity in all ceramic tableware.

It is impossible to estimate how much uranium-containing dinnerware is in circulation today. The Homer Laughlin Company has not traced all its ware, and a few other companies that produced similar dinnerware are no longer in business. In addition, although several studies point to the bright orange-red as emitting the most radioactivity, readings have been noted on other colors; they vary between companies and plate type, making it difficult to draw specific conclusions (a 1986 letter from the NRA to a consumer inquiry says that the easiest way to determine the amount of radioactivity in the dishes in question would be to obtain a Geiger counter).

Various studies led the FDA to conclude that the amount of radiation emitted by the tableware did not pose a health hazard. At the same time, the agency concluded that such exposures are "unnecessary," "clearly avoidable" and are "of no benefit to the public."

In 1981, the Buffalo office of the New York State Health Department told county health departments that the greater hazard with such ceramic tableware was the leaching of lead—which was used primarily to help stabilize the uranium in the glazes—into the food. In fact, leach studies performed on samples of several brands of uranium-glazed tableware by both the FDA and the New York State Health Department have shown that lead and uranium leach at levels above federal guidelines.

At Homer Laughlin, where the radioactivity issue has become folklore, the company has continually denied that its uranium-glazed dinnerware poses a health hazard to the public. As for the lead leaching, Ralph Franke, Director of Research at the company, says that the private lab which has been regularly analyzing its pottery since the early 1960s has always found the company's ware to be well below federal lead-leach levels.

Franke says that it became "such a hassle" to make Fiesta, with federal inspectors continually monitoring the facilities with Geiger counters, the adverse publicity and the general decline in the popularity of art deco, that Homer Laughlin discontinued making it in 1972. The company has, however, reintroduced its Fiesta line in an attempt to cash in on what they perceive to be an art deco comeback. With the exception of the rose shade, which contains a small amount of lead in the glaze, the re-issued Fiesta contains no lead or uranium.

Concern about radioactivity or lead notwithstanding, vintage ceramic tableware continues to be sought after by flea-market followers; the most expensive of the Fiesta line being none other than the mango-red made in the 1930s and 1940s. According to Sharon Huxford, co-author of The Collectors Encyclopedia of Fiesta, the red developed a status because of its popularity at the time and because the inclusion of uranium made the pieces more costly to begin with.[51]

Mercury

Sweden decided to ban the further use of mercury amalgam dental fillings in children and young adults, effective in June of 1995. It was further declared that all Swedish citizens would be protected from any further amalgams as of January 1997. The Swedish government actually does its own research on this subject, concluding that 250,000 Swedes had immune and other health disorders felt to be directly related to their amalgams. They stated the simple purpose of this ban was to protect the people and the environment. Even more recently, Denmark decided to ban amalgam, effective in January of 1999.

Approximately three years ago the German Health Ministry recommended to the German Dental Association that no further amalgam restorations be placed in children, pregnant women, and individuals with kidney disease. In December of 1993 this proposal was extended to include all women of child-bearing age, pregnant or not. The Association, in a most interesting response to the Ministry, then replied that if any further limitations on the placement of amalgams were suggested, it would simply have to advise its members to stop using amalgams completely due to the increasing chances of legal action being brought against any of them. Such legal concerns must have had Degussa, Germany's largest producer of amalgam and the world's largest producer of metals for dentistry, already in a similar frame of mind as that of the Dental Association, as it completely shut down its amalgam production. Degussa took the posture that it would reinitiate such production when mercury was proved to be safe in the body.

While not nearly as broad in scope as the above European initiatives, the United States did have Proposition 65 in California as a start. This was initially passed by California voters in 1986 to provide information to consumers on chemicals that can cause birth defects and reproductive problems. The Environmental Law Foundation in Oakland, California, in San Francisco County Superior Court on December 14, 1993, reached a settlement with Jeneric/Pentron Inc. of Wallingford, Connecticut (one of the nation's largest manufacturers and distributors of mercury amalgam dental filling material) after legally contesting the purported safety of amalgams.[52] In compliance with the spirit of Proposition 65, Jeneric agreed to send warning signs to all California dentists who purchase its mercury amalgam products. The warning signs, earmarked by an inverted yellow triangle, and to be displayed prominently in the dentists' offices, will state:

WARNING: This office uses amalgam filling materials which contain and expose you to mercury, a chemical known to the State of California to cause birth defects and other reproductive harm. Please consult your dentist for more information.[52, 53]

Recently, the California State Board of Dental Examiners, by unanimous vote, approved a two-page document entitled "Dental Materials Fact Sheet." The purported intent of this document is to encourage discussion between patient and dentist in the selection of dental materials best suited to the patient's dental health, and it will be made available to all licensed dentists in California. The Board agreed that elemental mercury is a toxic substance, and it acknowledged that research has shown that free mercury can escape from amalgam filling and be absorbed by the body. The document even states, "Some elements contained in composites have been determined to be cytotoxic and carcinogenic."[54]

It would be wonderful if amalgam was safe. It is, in fact, an inexpensive and durable substance whose properties allow for a technically quick and relatively easy placement in the mouth. And, as all its proponents are quick to point out, it has been in use in this capacity for over 150 years now. Were it not so cheap and easy to use, its profoundly toxic effects on general health would have been very apparent long ago, but since such a large percentage of the civilized population have them in place, there has been no control population with which to readily compare differences in health.

In the early 1800s the National Association of Dental Surgeons actually advocated the elimination of mercury amalgam, but its cheapness kept many dentists using it in spite of its toxicity. This Association disbanded several decades later, and the precursor to today's American Dental Association (ADA), the National Dental Association, came into being, proclaiming amalgam's safety, although this was just a political statement then, as it is now.

For some time, it was simply asserted by the ADA that the mercury amalgam (composed of approximately 50 percent mercury, along with copper, tin, silver, and zinc) was a tightly bound chemical complex that would not permit any leakage or release of mercury. This was proved conclusively wrong by Vimy and Lorscheider in 1985 when they demonstrated that the air inside the mouth with amalgams continually contained elemental mercury vapor, and the "dynamic of chewing increased this vapor level substantially."[55] They further concluded that the amount of mercury released daily in patients with 12 or more amalgams either exceeded or comprised a major percentage of the maximal permissible dose of mercury from all environmental sources, as established by the World Health Organization (WHO) in 1972 (although it is highly debatable whether a heavy metal as toxic as mercury should really have a politically derived, "permissible" dose). The most acceptable exposure would be the one most closely approximating zero. Gay et al. published similar conclusions on amalgam mercury leakage even earlier in the Lancet medical journal.[55, 56] Faced with this information, the ADA smoothly shifted gears and asserted that, although mercury was slowly released from the amalgams, the amounts were too small to matter, completely ignoring the significance of the data and findings of Vimy and Lorscheider.[55]

1. Mercury's Widespread Toxic Effects

Mercury is the most toxic (nonradioactive) inorganic heavy metal known to man. Its effects are enormously widespread and really leave no part or system of the body untouched. Exposure to mercury through its numerous industrial and commercial uses amounts for significant accumulation in our bodies. The added load arising from amalgams often tips the scales in favor of toxicity.

The previously noted elemental mercury vapor that emanates from amalgams is almost completely inhaled, little of it being lost outside of the mouth and body. Such inhalation allows for a rapid and complete absorption across the alveolar membrane in the lungs. This mercury easily crosses the blood/brain barrier (the brain and nervous system's natural defense against many toxic substances), subsequently binding very strongly to the sulfur-containing proteins of the nervous tissue.

This same affinity for binding sulfur allows its deposit in virtually all of the body's other tissues and organs. In fact, the much-maligned scapegoat in today's health, cholesterol, appears to actually afford a protective mechanism against the slow and insidious release of mercury into the bloodstream by binding it up and allowing it to be excreted before it gets its grips into any of the body's tissues. High cholesterol levels may represent just a healthy metabolism doing its best to neutralize the continual release of a toxin.[57] Patients who undergo amalgam removal consistently show shifts of their cholesterol into or toward the normal range, often within days of such removal.

Investigators have noted that low cholesterol levels, or sudden drops in cholesterol, appear to cause an increase in the incidence of homicides, suicides, and accidents.[58] Sudden fits of uncontrolled anger and temper, severe depression, and loss of coordination and motor control are some of the most common manifestations of chronic mercury poisoning. Perhaps, then, cholesterol drops give the newly released mercury the "edge" in a body that already has significant mercury stored in its tissues. For those who may still doubt that mercury is really accumulating in their bodies from their amalgams, cadaver examinations have conclusively demonstrated that the greater the number of amalgams, the greater the amount of mercury found in the brain tissues. Just five amalgams increased brain mercury levels threefold over controls.[59]

2. Mercury and Pregnancy

Mercury is even less considerate to the unborn. Methylmercury, the organic form of mercury that forms after oral ingestion of mercury, is 100 times more toxic than the previously mentioned elemental mercury. This form of mercury quickly and easily passes the placental barrier and builds up to 30 percent higher red blood cell levels in the fetus than in the mother.[60]

Stillbirths are significantly correlated with maternal blood mercury levels, and, as might be expected, mothers with larger numbers of amalgams tended to have higher maternal blood mercury levels. Depending upon the degree of methylmercury exposure to the fetus, the damage rendered can range from death (stillbirth) to mental retardation to an apparently normal birth, but sometimes such seeming normalcy at birth is followed by psychomotor and behavioral disturbances as the nervous system attempts to mature in the growing child. Such disturbed children had significant increases in their mercury and lead levels. Even after birth, the blood mercury levels were higher in the infants than in their mothers for the first four months. Furthermore, these elevated levels were supplemented by the mercury transmitted through breast-feeding.[61]

B. Official ADA Stand

Presently, in the ADA "Code of Professional Conduct," it states:

> Based on available scientific data, the ADA has determined through the adoption of Resolution 42H-1986 that the removal of amalgam restor-ation from the non-allergic patient for the alleged purpose of removing toxic substances from the body, when such treatment is

performed solely at the recommendation or suggestion
of the dentist, is improper and unethical.

In other words, the ADA is telling the dentists of America that they
don't have the right to counsel their patients regarding the poisonous
effects of mercury, unless, of course, they don't mind losing their
licenses to practice dentistry. Patient beware! Unless you are an
active advocate for the health of yourself and/or your family,
constantly researching the medical facts for himself, you should not
expect that your dentist to share the truth about mercury without being
asked to do so. After all, the dentist has to feed his family; his talents
aren't really directly applicable to much of anything else.

Sadly enough, more than a third of dentists in a survey published
in the December 1989 issue of Dentist magazine believe that all silver
(mercury) alloy fillings should be removed and replaced with
composites. Exactly what the ADA's true intents and goals are is hard
to fathom, but it's hard to conceive that patient welfare is very high on
their agenda, if, indeed, it's on the agenda at all.

After CBS-TV ran a "60 Minutes" segment on the amalgam issue
in December of 1990, the Washington State Dental Association,
amazingly, promptly informed its members that their patients did not
have a right to know that their "silver fillings" contained mercury.[62]
Moreover, that segment purportedly received the highest viewer
response ever, but it's never been repeated, despite the fact that other
episodes are frequently rerun. Is there another unknown agenda
here?

In April 1994, the Journal of the American Dental Association, in
an analysis of a review of the benefits and risks of dental amalgam
conducted by the U.S. Public Health Service (USPHS) and published
in 1993, made prominent note of the estimated costs for replacement
of dental amalgams in the entire country. It was stated that one-time
replacement of all existing amalgams in permanent posterior teeth
would cost 248 billion dollars. It was further pointed out that the
increased cost in 1990 had alternative restorative materials been used

instead of the amalgams that were placed in the 96 million treated teeth would have been 12.4 billion dollars.[63] Money usually talks, and that's a lot of money.

C. Multiple Sclerosis—Predominantly a Dental Disease?

In his classic original description of multiple sclerosis (MS) around the mid-1830s, Cruveilhier attributed the disease to suppression of sweat, and according to the Fifth Edition of the Principles of Neurology (1993) textbook, "since that time there has been endless speculation about the etiology."[64] It is very interesting to note that the first mercury fillings were placed in unwitting mouths in France shortly before Cruveilhier's observation. Around 1826, M. Taveau in Paris began promoting simple silver/mercury paste fillings. Additional amalgam components to this paste followed shortly thereafter when patients consistently demonstrated fractured teeth due to expansion of the paste after setting.[65] For the first reported appearance of MS to appear only a very few years after the first insertions of dental mercury should be hard to completely ignore for even the most ardent of the remaining amalgam advocates. Furthermore, Cruveilhier's initial observation that the disease seemed to relate to sweat suppression actually meshes nicely with today's knowledge that sweat induction (as in a dry sauna) is still one of the best ways to eliminate mercury from the body's stores.[18]

Early in the course of the disease, and often when the diagnosis is not yet secure, MS will have characteristic fluctuations in symptom severity, but when the motor (muscle) weakness progresses to the point of requiring a wheelchair, remission is all but out of the question, unless amalgam removal is undertaken. Dr. Hal Huggins, a dentist in Colorado Springs, Colorado, has been consistently witnessing improvement in MS patients undergoing amalgam removal for many years now, seeing clear symptomatic and laboratory test improvement in 80-85 percent of them presently, and often even seeing wheelchair patients who had not been wheelchair-bound for too long walk again.[68] If these results are to be written off as "anecdotal" or "placebo effect," then Dr. Huggins certainly has the largest collection of sustained recurring anecdotal placebo responses of MS patients to amalgam removal in the world today. Witnessing such a response firsthand in a friend or relative is all the scientific literature that most people, including medical professionals, need.

1. Your Mouth Is Wired

The brain and central nervous system (CNS) are also strongly affected by the electrical current present in all mouths containing metal. This phenomenon is called oral galvanism. These currents can be measured very easily with a probe and a microammeter. Amalgams, metallic crowns, and braces generally all register from 1 to 100 microamperes of current in a positive or negative polarity. The natural currents found in the brain are in the range of 7 to 9 nanoamperes, making the mouth currents anywhere from 100 to 10,000 times more powerful (and the base of the brain is roughly only an inch away from the upper teeth).[66] Small wonder, then, that so many MS and miscellaneous neurological patients will demonstrate an immediate improvement in clinical status, often manifest as improved muscle strength and coordination. Similarly, removal of highly electrical dental material has shown occasional immediate effects on diverse, other symptoms such as severe migraine headaches, chronic cough, jaw pain, muscle cramping, chest pain, energy level, and even depression.

It is also very important to note at this juncture that when Dr. Huggins first began removing mercury amalgams in 1973, he was only successful in improving the clinical status and abnormal laboratory findings in approximately 10 percent of MS patients. His success rate reached about 60 percent in 1979 when he realized the importance of sequential removal of the amalgams according to the amount of current measured on each one, removing the ones with the highest negative current first.

About this time as well he realized the paramount importance of proper nutrition and supplementation of vitamins and minerals specifically based on each patient's laboratory profile. It was unfortunately too common for a patient to resolve almost completely clinically, later disregard the dietary and supplementation recommendations, and come crashing back to his pre-amalgam removal clinical status months or even years later. Even after amalgam removal, it was clear that all of the MS patients, even the ones with apparent complete recovery, would be walking a tightrope for the remainder of their lives, due to their remaining high total body stores of mercury and the frequently noted secondary immune reactions that were seen after amalgam removal, debilitating these patients whenever their bodies encountered more mercury, either environmentally or through ingestion, as in fish and seafood (foods extremely high in methyl-mercury content).

Other non-mercuric toxins need to be avoided as well to maintain an acceptable clinical status. As an example, fluoride in all its forms needs to be scrupulously avoided by such patients, as it can also retard clinical progress or even promote frank clinical relapse. Huggins himself is all too keenly aware of the nuances and persistence of MS as a disease, as he has had it for many years now, but keeps the symptoms largely in check by following the lifestyle modifications, diet, and supplementation regimens strictly. Finally, in the mid-to-late 1980s Dr. Huggins came to appreciate the importance of chronic dental infections, as seen in virtually all root canals and cavitations (healed over holes at the sites of previous extractions). When these teeth were removed and all new and old extraction sites properly cleaned, he finally reached his present success rate with MS.[68]

In a poll of 1,320 patients at Huggins Diagnostic Center (HDC), unexplained irritability, frequent depression, numbness and tingling of the extremities, chronic fatigue, tremors and difficulty with memory were seen in a majority. These symptoms are also among the most common symptoms in MS patients. Even when frank MS is not present, these symptoms in isolation all respond as well or better than MS with amalgam removal. It would appear logical to assume that many such patients with isolated symptoms as above could be at one end of a continuum that will lead to MS or a similar debilitating neurological syndrome if the continual absorption of mercury into their bodies is not addressed.

Over approximately 20 years, Dr. Huggins has treated literally thousands of patients. In recent years; however, he's noticed that his MS patients are younger, and the disease is progressing more rapidly. Many patients become absolutely wheelchair-bound even when their first clear symptoms were only 2 to 3 years earlier. Around the same time this was observed, the "high-copper" amalgam (an amalgam with a substantially greater amount of copper) began to be utilized with increasing frequency. This amalgam was found to release 50 times more mercury than the previous conventional amalgam.[67] It seems the more dentistry advanced, the worse it got. You might also be amazed (and, perhaps, stupefied) to know that high-copper amalgams were actually outlawed several decades ago due to their severe cytotoxicity (e.g., their ability to kill cells in the body).[68]

D. More Toxic Effects

In studies where test animals inhaled mercury vapor, the uptake of mercury was greatest in the kidneys, followed by brain, heart, intestine, and liver in decreasing order.[69] Additional target sites include the testes, ovaries, and pancreas, as well as the thyroid, adrenal, and pituitary glands. Essential hypertension (high blood pressure) and heart disease have showed steady increases in recent decades, and chronic mercury toxicity can be related to them both. Smaller amounts of inorganic mercury elevate the blood pressure, and larger amounts can cause direct heart muscle damage (cardiomyopathy), resulting in heart failure and an ultimate lowering of blood pressure.[70] In a very recent study published in the Journal of the American College of Cardiology, nearly half of 673 patients with enlarged hearts and heart failure were classified as idiopathic, or cause unknown, despite extensive testing.[71] Of extreme interest in this regard is the discovery by Soviet researchers that mercury binds avidly to sulfur containing contractile (squeezing ability) protein sites in the heart muscle itself, a property that could ultimately cause the poorly functioning, enlarged hearts as mentioned above. Perhaps that "cause unknown" fraction of the patients is not completely unknown after all. Haber and others at the Mount Sinai School of Medicine found that in patients with chronic heart failure, heart function was clearly worse in those patients with the lower cholesterols.[72] It would seem that the lower cholesterols made a toxic situation even more toxic. And, in addition to being a possible primary cause for many cases of idiopathic heart failure, mercury toxicity could very readily serve as a co-factor in worsening heart failure in cases that have been assigned another etiology, or cause.

In the pancreas there are groups of cells called the Islets of Langerhans, which function to secrete the body's insulin and thereby regulate the blood sugar. Mercury has an affinity for these cells and appears to directly affect the sugar metabolism in some patients.[73] In fact, diabetics requiring insulin injections who undergo amalgam removal frequently show a subsequent decreased need for insulin. When this fact is ignored in the course of amalgam removal, some patients can go into insulin shock (severe low blood sugar) when their blood sugars are not monitored closely and their insulin doses are not appropriately lowered.

Autoimmune and collagen vascular diseases are also commonly caused, or worsened by, chronic mercury poisoning. The disease known to the public as "Lupus" is probably the most infamous of this disease group. These diseases are characterized by a duping of the body's immune system, in which the body actually attacks certain of its own tissues after they sustain damage from outside agencies or toxins, such as mercury. Autoantibodies, produced by the body and directed against itself, allow for the laboratory diagnosis of these diseases. In experimental animals, mercury exposure induced such autoantibody production in greater than 90 percent of the time.[74]

1. Curing the Incurable?

Leukemias, as well as other malignancies, have also often been observed to respond to the Huggins protocol. A particularly interesting leukemia, and still totally incurable by traditional methods, is chronic lymphocytic leukemia. A very slowly progressive disease, and often asymptomatic for years in the older patients it commonly involves, it is characterized by elevations, sometimes astronomical, of the white blood cells. These same cells have been noted to increase after amalgam placement. Amalgam removal has been accompanied by rapid drops in this cell count in some leukemics after only a few days. Almost ironically, some leukemia patients actually seem to be mounting an excessive immune response to the mercury exposure, and this excessive response is what eventually kills them.[76]

Other "incurable" diseases have also responded to total dental revision. These diseases include amyotrophic lateral sclerosis (ALS, or "Lou Gehrig's Disease"), Parkinson's disease, and even Alzheimer's disease. It should be emphasized, however, that although these diseases do often respond, the clinical improvements are generally not as profound or as quick to appear as in some other diseases such as chronic fatigue syndrome or MS. Bear in mind, however, that modern medicine is presently doing very little for all three, beyond diagnosis and custodial care.

2. Deadly Alternatives

Unfortunately, from a dental perspective, maintaining good health involves more than just avoidance of mercury. Nickel is rapidly gaining a severely toxic reputation as well. Most partial dentures are made of nickel. Approximately 80 percent of crowns utilize nickel, even so-called porcelain crowns, where nickel is often the base on which the porcelain is fired. The braces worn by many children and young adults are usually nickel. (Stainless steel is usually nickel alloy.) Nickel compounds have been unequivocally implicated as human respiratory carcinogens in epidemic-logical studies of nickel refinery workers.[75] Moreover, recent data is indicating a relationship between nickel crowns and breast cancer in women.[76] Although many children appear to tolerate braces without apparent difficulty, many children also become chronically ill or lose mental acuity in school, and no thought is ever given to the nickel braces as an etiology. For those parents who must put braces into their children's mouths, or even nickel crowns, consideration should be given to pre and post insertion checks on lymphocyte viability, a measure of the percentage of lymphocytes (a white blood cell important in the immune process) that are actually alive and functional. Nickel is notorious for killing these important cells, and a significant drop (>10 percent or more from a normal of 95-100 percent alive) is a red flag, indicative of an immune system virtually screaming to get the nickel out.[75]

Cast glass crowns and inlays are some of the latest dental materials. However, be aware that this "glass" generally contains more than 25 percent aluminum as aluminum oxide. About 80 percent of patients with such dental work show laboratory findings consistent with some drop in immune function, in spite of the propaganda fed dentists that all of the components in this "glass" remain tightly bound and safe (just like the reassurances initially that amalgam remained forever intact).

Composites, filling materials consisting of ground glass powder mixed with a plastic binder, demonstrate laboratory immune reactivity in patients about 50 percent of the time. Certain composites (they are not all the same) evoked this reactivity in over 90 percent of patients.[68] Composites certainly can be good as long as the individual patient has had the appropriate testing to verify a low level of immune reactivity to the components.

3. Serum Compatibility Testing

Presently, there are over 1000 different metal alloys from which to choose. It was found that the only safe (and scientific) way to know what was best (and least toxic) in any given mouth was to check the immune reactivity of the patient's blood serum against all of the commonly used dental materials. Virtually everybody shows some reactivity to all materials, but the degrees of reactivity can range widely, and such a laboratory profiling allows for an intelligent selection of the least reactive dental materials in a given patient. Some unfortunate patients who were aware only of mercury's toxicity have had wonderful recoveries from their various illnesses after amalgam removal, only to precipitously relapse after reacting severely to replacement dental fillings. Certainly, some composites used are usually safe, but is avoiding a blood test worth one's health? There is no excuse not to use such testing prior to dental work, especially in the case of children preparing for their first dental visits.

E. Hidden Dental Infections—Mercury's Little Brother

Dr. Weston A. Price, in 1923, published his 1100 page, two volume treatise on dental infections, entitled *Dental Infections, Oral and Systemic* (Vol. 1) and *Dental Infections and the Degenerative Diseases* (Vol. 2). The results of his research were nothing short of profound. While it may seem a ridiculous statement to make, he found that there seemed to be hardly any disease or disease process that was not either primarily caused by dental infections or just worsened by them. The heart and circulatory system appeared to be favorite target sites for the bacteria and/or their toxins. Dr. Price observed angina pectoris, phlebitis, hypertension, heart block, anemia, and inflammation of the heart muscle to often be side effects of root canal therapy. He also reported that he would sometimes see heart patients with outwardly normal appearing root canal teeth resolve most or all of their symptoms upon removal of those teeth.

One of Dr. Price's experimental techniques was to implant extracted root canal teeth under the skin of rabbits after removal from patients with various illnesses. Amazingly, not only would the rabbit become ill, but the animal would reliably develop precisely the same primary disease that the human tooth donor had. In this fashion, Dr. Price transferred to rabbits arthritis; heart lesions; kidney, liver, and gall bladder disease; anemia; pneumonia; appendicitis; eye, ear, and skin disorders; and nervous system disorders. It also should be noted that he would use the same tooth, or shavings from it, sequentially in a minimum of 30 rabbits, consistently demonstrating the same disease transferal.

The underlying premise behind the toxicity of root canals is basically twofold: (1) they cannot be sterilized, and they all demonstrate chronic infection, however normal they may appear clinically, and (2) the anaerobic (oxygen-lacking) environment they offer in the root tips allows otherwise benign, oxygen-utilizing mouth bacteria to undergo a toxic transformation, causing the production of thioethers as a bacterial byproduct. These thioethers alone proved to be potent sources of disease. Dr. Price would grind up the root canal teeth, wash them with a solution, filter the solution through a bacteria-retaining filter, and upon injecting the germ-free solution (containing the thioethers), still cause the same reactions in the rabbits as with the whole teeth.

When Dr. Huggins discovered Dr. Price's research, noting especially the amazing results reported in numerous patients with all variety of chronic degenerative diseases, he incorporated root canal extraction into his routine of dental revision, and he immediately saw a dramatic further increase in positive patient response, as noted earlier. However, some patients would show an initial dramatic improvement for a few months, and then they would demonstrate a slow but progressive backslide, sometimes back to baseline, even when they were following all lifestyle, dietary, and supplement recommendations.[68]

Dr. Huggins then had what proved to be a truly brilliant insight. Every extracted tooth has its own capsule in the jawbone called the periodontal ligament, which was always routinely left behind after extractions (and, unfortunately, nearly all dentists still leave it in today). Dr. Huggins reasoned that the deep-seated root canal infections likely infected this ligament as well, and once the bone healed over top, the same circumstances of chronic infection could persist even without the root canal tooth being any longer in place. He then initiated a quick and simple routing out of this ligament, along with about 1 millimeter of jawbone following extractions. Similarly, because the ligaments were always left behind, he postulated that most people probably never filled bone in completely, even after decades-old extractions (often the case with wisdom teeth removal), and upon routine exploration these old sites did usually show a hole, or cavitation. These sites were then approached like fresh extractions, with removal of the old ligaments and some surrounding bone, and they would nearly always show a prompt healing, this time with a totally filling in of bone.

One final extension of this reasoning led Dr. Huggins to realize that even previously normal teeth that were severely traumatized with loss of nerve and blood supply ("dead teeth") would then usually become infected eventually just like root canal teeth, needing extraction with site cleaning for optimal results. These final modifications to Dr. Huggins' approach to dental revision have brought him to his present level of success.[68]

Huggins makes an additional note about edentulous (toothless) patients, commenting that most such patients probably have numerous cavitations, since the periodontal ligaments were likely not removed at the time of the extractions. In the older patients lacking a robust immune system such a situation could still severely compromise their health in any of the numerous ways already mentioned, even in the absence of any implanted metals or other dental materials. Note should also be made of the fact that the "pink" in the artificial denture plates comes from mercury or cadmium-containing pigment. Clear plates with just a few front teeth separated by the pink component is often a reasonable compromise between cosmetic appeal and additional mercury or cadmium exposure.

Huggins has also observed the direct effects of various dietary and supple-mentation interventions on laboratory values, giving him clear feedback on the foods best to avoid and the vitamins and minerals best to take.[68] If you have a mouth full of amalgams and/or root canals, and you feel that you may well have compromised health because of them, a good first step to recovery is to obtain your own lab work as a guide to what your vitamin and mineral supplementation should be.

Many people have excess stores of certain minerals, and a "shotgun" approach to supplementation with many multivitamin/multimineral formulations can worsen a situation rather than help, making some of the excess minerals even more excessive and thereby toxic. For example, the body does require some copper and chromium, but these can easily be made toxic if their levels are already somewhat high and any supplements containing these minerals are taken. Similarly, if such levels are deficient and not specifically addressed, they tend to remain deficient.

Some Common Non-Dental Sources of Mercury Exposure

Food
* Grains treated with methyl mercury fungicides (wheat).
* Kelp and other seaweeds.
* Large saltwater fish like swordfish, salmon, cod, etc.
* Shellfish, including shrimp, lobster, crab, oyster, etc.
* Tuna (canned or fresh)

Cosmetics
* Clairol hair dye.
* Mascara, especially waterproof.
* Skin-lightening creams.

Industrial Processes
There are numerous industries in which workers can be exposed to mercury.

Some of the most mercury-toxic processes in the working world include:

* Alloy handling
* Bactericide production
* Battery manufacturing
* Carbon brush making
* Artificial silk manufacturing
* Barometer manufacturing
* Bronzing
* Caustic soda production

* Disinfectant production
* Dye manufacturing
* Electroplating
* Farming
* Fingerprint detecting
* Fungicide production
* Gold and silver extracting
* Insecticide production
* Laboratory work (chemical)
* Lead-mercury soldering
* Mining
* Neon light manufacturing
* Printing
* Photography
* Steel etching
* Textile printing
* Vinyl chloride manufacturing
* Drug manufacturing
* Electric apparatus manufacturing
* Explosives production
* Felt manufacturing
* Fish cannery work
* Fur processing
* Ink making
* Jewelry production and repair
* Lampmaking (fluorescent)
* Manometer manufacturing
* Mirror production
* Paint manufacturing
* Papermaking
* Seed handling
* Tannery work
* Thermometer making
* Working with mercury

1. What Should One Do?

DO NOT get amalgams removed by a dentist who has not convinced you of his awareness of mercury toxicity, or of his ability to perform such removals with maximal safety precautions. Although the numbers of such dentists are very slowly increasing, the numbers are still very small for the vast number of patients that need attention. Your health could be worsened rather than improved if too much mercury is dumped into your body during the removal process.

Probably the first consideration in this process is an honest evaluation of how healthy you really are. If you are robust, full of energy, and free of any significant symptoms, then rushing out and getting a complete dental revision may not be the way to go, especially if you are one of the many who are financially compromised and only able to address the usual bills. This should not dissuade you from considering adopting certain significant dietary modifications and undergoing minimal laboratory testing to determine a program of specific supplementation, as mentioned above. Everyone should undergo full biocompatibility testing to determine their individual sensitivities prior to further (or first) dental work. Prevention is always so much easier than undoing what has already been done. If your health is severely compromised, and especially if you can trace the biggest, or most rapid, decline in your health to be close in time to your last dental intervention, then a complete dental revision may be the most important factor in regaining or improving your health. Minimal protocol standards for amalgam removal would include the following:

—Use of a "rubber dam" (to prevent swallowing of amalgam).

—Sequential removal of fillings based on electrical measurements.

—Appointment scheduling to avoid the weekly drops in immune system strength.

—Operation room air filtration to remove air-borne mercury from the drilling.

—Serum compatibility testing to determine replacement filling materials.

—Appropriate laboratory testing to direct dietary changes and appropriate supplementation regimens.

—In sicker patients, intravenous vitamin C and possibly even chelation infusions during the removal procedures, if at all possible (to facilitate mercury processing and excretion).

2. Non-Dental Mercury Exposure

In spite of its severe toxicity, mercury is just about everywhere. This is especially important to know for the patient who has recovered or improved from a mercury-toxic condition after amalgam removal and wants to maintain improvement. Mercury has been used by man for at least 2,000 years, and more than 60 occupations involve mercury exposure. These include manufacture of pesticides, insecticides, and fungicides; manufacture of mercury-containing instruments, lamps, neon lights, batteries, paper, paint, dye, electrical equipment, and jewelry; and dentistry.[77]

The table listing the multiple places where mercury can be found even more clearly demonstrates its widespread use. Many cosmetics and medicinals still contain mercury. Many eyedrops are still sterilized by thimerasol, a mercury-containing preservative. Many vaccines are similarly sterilized.

Lead Production and Manufacturing

The use of lead predates the Christian era by 2,000 to 4,000 years. The ancients employed the metal in coinage, weights, piping, corrosion-resistant containers, ceramic glazes, and in glass. Man still uses lead for most of these same products. In addition, a host of newer uses has been developed, including use in storage batteries, bearings, cable coverings, ammunition, type metal, pigments, high octane gasoline, and radiation shields.

The United States has been the world's leading producer of lead since 1893 and its chief consumer since about 1900. From 1925 through 1958 domestic mines produced 14.5 million tons of lead, or one-quarter of the world mine output. In this period U.S. industries consumed about 40 percent of the world supply of new lead. World War II marked the beginning of U.S. dependence on imports for a large portion of its needs. For many years the United States had been self-sufficient in lead, but beginning in 1940 and continuing thereafter, greatly increased domestic requirements necessitated the importation of large tonnages of foreign metal, ores and concentrates to augment home supplies. Australia, Mexico, and Canada are the second, third, and fourth largest producers of lead, and historically Mexico and Canada have been the chief sources of U.S. imports. Major producers of lead on a mine basis, in order of output in 1958, were Australia, U.S.S.R., United States, Mexico, Canada, Peru, Morocco, Yugoslavia, and Southwest Africa. Together with the United States, these eight countries produced almost three-fourths of the total world output.[87]

The total lead content of measured and indicated reserves of lead ore throughout the world is estimated to be approximately 49 million tons. Of this, 14.6 million tons is in North America, 2.5 million in South America, 13.7 million in Europe, 3.5 million in Africa, 2 million in Asia, and 12.5 million in Australia. North American reserves are principally in Canada. The United States has a reserve of 2.9 million tons of lead and that of Mexico is estimated to be about 3.5 million tons.

Although several hundred firms are engaged in one or more aspects of the lead industry, fewer than a dozen dominate it. The five largest producers normally supply over half of the total U.S. mine output; almost all the primary refining capacity is controlled by four companies; two firms process, recover, and market about one-half the secondary lead, and four companies are preeminent in marketing lead.

Consumption of lead is expected to increase in the United States but at a lower rate than in the rest of the world. Despite development of numerous new uses for lead, the effect probably will be offset by the use of substitute materials. World reserves of lead seem adequate for many decades. It is anticipated that annual production of both primary and secondary lead in the United States will remain relatively constant with increased requirements being met by increased imports of metal.[87]

Oversupply and low prices form a real but short-term problem which tends to be self correcting. The low prices stimulate consumption, which reduces stocks to stimulate price to a level consonant with demand. The major and long-term problems of the industry are those associated with maintaining an ample supply of metal at prices that stimulate use. They include economic ore discovery, and the solution of numerous mining, milling, and smelting problems, which are essential to conservation and a low cost per unit of output.[87]

A. Background

Lead was mined and smelted in Virginia as early as 1621, and the discovery of lead in the Upper Mississippi Valley was reported in

1690. The shallow mines of the Upper Mississippi Valley were important sources of lead during much of the 19th century. In 1867, discoveries in southeast Missouri at depths of over 100 feet led to the further development there of one of the two most productive lead regions in the world. The year 1958 marked the 51st consecutive year in which southeastern Missouri was the leading lead-producing mining district in the United States.

Lead production in the United States was annually about 1,000 tons from 1801 through 1810, but increased progressively to about 13,000 tons of lead a year from 1831 to 1840, approximately 10 percent of world output. The Missouri and Upper Mississippi Valley areas were the principal producing regions. Lead production from 1840 through 1870 was essentially at the same rate as in the decade 1831-1840.

Completion of the first transcontinental railway in 1869 gave further impetus to the lead-mining industry. An era of prospecting followed that led to discovery of lead deposits at Eureka, Nevada; Bingham Canyon, Park City, and Big and Little Cottonwood Canyons, Utah; Leadville, Colorado; Cerro Gordo, California; Bonne Terre and Joplin, Missouri; and in the Coeur d'Alene, Idaho region. Production increased sharply in the 1870's, and both mine and smelter production grew rapidly during the 1880's and 1890's. The United States became the world's largest lead producer in 1893, and during 1891-1900 production annually exceeded 200,000 tons, or more than 25 percent of the total world output. Production increased remarkably from 1900 through 1929. In 1900, the mine production of recoverable lead totaled 270,000 tons, and by 1925 it was 684,000 tons—more than one-third of total world output.[87]

Many new techniques to improve mining, milling, and smelting were introduced or gained wide acceptance in this period. These techniques included mechanized haulage underground, use of centrifugal pumps, widespread use of froth flotation (including differential flotation) and the development of centering devices to improve the structure and smelting qualities of the blast furnace feed. In the latter part of this period, recovery from scrap became the important element of lead supply that it still is, comprising over a third of the U.S. total lead supply.

The United States has continued to be one of the world's leading lead-mining nations since 1929; but owing to increased requirements, beginning in 1940, imports have been necessary to fulfill domestic needs.[87]

1. Size, Organization and Geographic Distribution of the Industry

Lead is one of the most widely used metals. It has ranked fifth in tonnage of metals produced and consumed in the world, following steel, copper, aluminum, and zinc, in that order. The United States was the world's foremost producer of lead as ore and as refined metal from 1893 through 1956. In 1957 Australia gained first place in mine production and in 1958 Australia and U.S.S.R. both surpassed the United States; the United States has kept ahead in production of refined metal. The countries of the Western Hemisphere produced 36 percent of the 1958 mine output of the world, with the refined lead output of the principal six countries—United States, U.S.S.R., Australia, Mexico, West Germany, and Canada—accounting for 64 percent of the world output, 38 percent in the Western Hemisphere.[87]

In the U.S., the lead industry is primarily composed of about 500 companies engaged in mining, smelting and refining, and marketing lead. In mining, seven companies—the St. Joseph Lead Company, United States Smelting, Refining and Mining Company, The Bunker Hill Company, American Smelting and Refining Company, The Eagle-Picher Company, The Anaconda Company, and National Lead Company—produce more than 65 percent of the domestic output. The first six of these companies own the twelve primary lead smelters and refineries in the United States; the National Lead Company, with American Smelting and Refining Company, operates secondary smelters having almost 50 percent of the total U.S. secondary capacity. The same seven firms, together with American Metal Climax Inc. and C. Tennant Sons and Company (sales agents for important foreign producers), are leaders in competitive lead marketing, and all these companies, except United States Smelting, Refining and Mining, Bunker Hill, Anaconda, C. Tennant Sons and Company, have lead mines abroad.

The secondary lead industry constitutes an integral part of the lead industry. Scrap lead is processed at both primary and secondary smelting plants to yield refined lead, antimonial lead and various lead alloys. In all, about 200 companies melt and smelt lead scrap, and the total lead recovered from scrap materials has, for the past 13 years, exceeded domestic mine production and, in many years, exceeded production of primary refined lead from both domestic and imported ones.[87]

About 97 percent of domestic mine lead output is from the area west of the Mississippi River. Nearly 60 percent of the lead mine production of the United States since 1875 has been from the western states, chiefly Idaho, Utah, Colorado, Montana, Washington, Arizona, and California. The most important western districts or regions are the Coeur d'Alene, Idaho; West Mountain, Utah; Summit Valley (Butte), Montana; Metaline, Washington; Coso (Darwin), California; and Upper San Miguel, Colorado.

The remaining mine production comes predominantly from the west central states with minor amounts from the eastern states. The southeastern Missouri lead district has been the leading producer of lead in the United States for 51 consecutive years. In 1958 output from this district totaled 113,000 short tons or more than double that of the Coeur d'Alene region, Idaho, which was the second most important lead district.[87]

2. Definition of Terms, Grades, and Specifications

Lead is a metal element, chemical symbol Pb, group IV and series 9 of the periodic system, atomic number 82, chemical atomic weight 207.21, stable isotopes 204, 206, 207, and 208, radioactive isotopes 209, 210, 211, and 214. It is a soft, heavy metal, malleable, but only slightly ductile. The specific gravity is 11.35 for cast lead at 20 degrees C. The melting point is 327.4o C and the boiling point about 1,700 degrees C at 760 mm pressure. Lead is the ultimate decomposition product of certain radioactive disintegrations; that from uranium has an atomic weight of about 206; that from thorium, 208. Refined lead is marketed for use in the following seven grades: Corroding, chemical, acid, copper lead, common desilverized A, common desilverized B, and soft undesilverized.

Large quantities of lead are marketed in various lead and copper-base alloys. The lead-base alloys are roughly classified into bearing alloys, solders, type metals, low melting alloys, and the numerous antimony or hard lead alloys. The copper-base alloys containing lead are various grades of leaded brasses and leaded bronzes.

Lead scrap is lead in metal or refuse form that lacks utility and must be remelted or smelted and brought to consumer standard by refining or alloying. The only scrap reported by the Bureau of Mines is that shipped or sold. Home scrap is processed at the plant of generation to secondary metal and is not an item of commerce nor included in lead statistics.

B. Technology
    1. Geology
    The common lead minerals are galena (lead sulfide), cerussite (lead carbonate), and anglesite (lead sulfate). Galena is the most abundant lead mineral found in deposits that have been exploited in the United States. Galena is commonly associated with zinc, silver, gold, and iron minerals. However, in a few districts the ore is characterized by simple mineralization, with lead minerals present to the virtual exclusion of other ore minerals. A noteworthy example is southeastern Missouri, where the lead ores are usually composed of galena in an essentially nonmetallic gangue, with relatively little silver, zinc, copper, or other valuable metals present.

The more important economic deposits in the United States occur either as cavity fillings or replacements. The origin of the mineralization in the cavity filling and replacement deposits is similar. The theory that the mineral-bearing solutions were derived from a deep-seated igneous mass is most widely accepted today. Examples of the cavity-filling type of deposit are the San Juan, Colorado and the Upper Mississippi Valley districts. Replacement-type deposits are classified further as follows: Massive type, as at Leadville, Colorado, and Bingham and Tintic, Utah; lodes, as at Park City, Utah, and the Coeur d'Alene region of Idaho; disseminated, as in the Tri-State district and in southeast Missouri; and metasomatic, as represented by the central district of New Mexico.

2. Prospecting and Exploration

Lead prospecting and exploration were relatively simple in the 18th and 19th centuries, when such important districts as southeast Missouri, the Tri-State, Upper Mississippi Valley, and Austinville, Virginia were first brought into production. As the easily discovered ore bodies were depleted, it became necessary to employ new prospecting techniques involving careful geologic mapping of rock formations and structural features that control, or may control, ore deposition. Through intensive study the mining geologist and exploration engineer have increased their knowledge of the fundamentals of replacement, structural control, contact metamorphism, zonal distribution, structural barriers, wall-rock alteration, oxidation, and tectonics both before and after ore deposition and are utilizing this knowledge in guiding prospecting and exploration.

Utilizing geophysical[78, 79] and geochemical[80] techniques, largely pioneered in petroleum exploration, geologists and engineers have worked intensively seeking anomalous conditions indicative of conditions favorable to economic mineralization. These techniques have recently been used with considerable success to find anomalies that lead to ore discoveries in the Little River area of New Brunswick, Manitouwadge area of Ontario, and southeast Missouri.

3. Mining

Underground methods employing either open or supported stopes are used in most lead mines, but open-pit methods have found limited application in the Tri-State district.[81] Other changes in equipment that have done much to improve mine output per man-shift include electric cap and floodlights at working faces, improved ventilation (including air conditioning), more efficient pumps, and powerful, lightweight rock drills using alloy steel or tungsten carbide bits.

4. Milling

Simple lead ores, such as coarsely disseminated lead or zinc-lead minerals occurring with a low-specific gravity gangue, are treated in heavy-medium equipment, jigs, and on tables after being crushed and rolled in closed circuit with screens or classifiers to give properly sized feed. Bulk or differential flotation of the slime products or of a re-ground middling product completes the flowsheet.[82] Ores of this kind are common in the mines of the Mississippi Valley and the eastern part of the United States, but in some instances the ores are concentrated wholly by flotation.

Complex sulfide ores such as those of the western United States consist of disseminated mixtures of finegrained lead and zinc sulfides, usually accompanied by pyrite, copper sulfides, and gold and silver in a quartz or quartz-calcite gangue. Such ores may be complicated further by partial oxidation of the sulfides and presence of high-specific-gravity gangue minerals, such as barite, siderite, and rhodochrosite. The usual procedure on such an ore is to crush and grind in closed circuit with classifying equipment to a size at which the ore minerals are freed from the gangue minerals. When the ore minerals are interlocked, the usual practice is to make a bulk sulfide concentrate, followed by regrinding and selective flotation.

Low capital and operating costs have extended the field of the sinkfloat method to include pretreatment of certain ores, permitting upgrading of ores diluted by non-selective mining methods. Present milling practices result in recovery of 85 to 94 percent of the sulfide lead and up to about 88 percent of the oxidized lead. Sulfide losses consist largely of the extremely fine particles, and although much research has been done and improvements have been made in the recovery of fines, the results are still unsatisfactory.

5. Smelting

Lead is recovered from its ores almost exclusively by smelting in blast furnaces or ore hearths employing carbon fuels. Ores or concentrates that contain few impurities may be reduced to metal in roasting hearths. Air is used to oxidize the sulfides and coke or coal to reduce the oxides. The ore hearth practice is followed at the Federal, Illinois and Galena, Kansas smelters where Newnam roasting hearths are used. The same process is employed at the New Broken Hill smelter of Rhodesia, but will shortly be replaced by blast-furnace methods.

In preparing the charge for blast-furnace smelting, the sulfur in the ores and concentrates is removed by a roasting-centering process, usually on a Dwight-Lloyd type centering hearth.[83] The centering equipment eliminates most of the sulfur and produces a blast-furnace feed of desirable characteristics from the mechanically mixed concentrate, undersized sinter particles, and byproduct dusts and fumes collected in the smelter. Normally the charge to the lead blast furnace consists of lead sinter, coke, fluxes (such as silica, lime, etc.), and some lead-bearing furnace by-products. Air is blown through the charge burning coke to CO and CO_2 and producing a temperature of approximately 1,400 degrees C. The carbon monoxide formed and the hot solid carbon reduce the oxidized lead compounds to bullion, which is tapped from the bottom or crucible of the furnace.

Any copper, iron, cobalt or nickel present in the ore combines with sulfur in the charge to form mattes which are tapped from the front of the blast furnace for subsequent treatment. Zinc present in the ore accumulates in the slag. When such slags contain 6 percent or more zinc, they are retreated in slag-fuming furnaces to recover the zinc and any remaining lead. The lead bullion contains precious metals present in the ore and metallic impurities which are recovered in the refining operation. The recent development by Imperial Smelting Corp., Ltd.

of a blast furnace process which recovers zinc and lead bullion simultaneously from mixed ores is of great interest.[84]

Lead recoveries at primary lead smelters are usually about 97-99 percent of the lead contained in the ore, offering a relatively small margin for improvement, except through lowering the unit cost of processing. The recent modernization programs center largely on charge preparation, roasting, dust collection, and material handling within the plants.[83] Such programs have more than doubled the productive capacity of lead blast furnaces since 1925.

6. Refining

Lead bullion from the Mississippi Valley ores, termed chemical lead, is pure enough for most commercial uses without further refining. Lead bullion produced from western and most foreign ore contains enough gold and silver to make extraction profitable. It also contains various base-metal impurities that must be removed before the lead is marketable for end use. The sequence of processes for softening and desilverizing lead bullion and the recovery of by-products is subject to many variations in practice.

Softening consists of the removal of copper, tin, antimony, and arsenic in a drossing or refining kettle. The copper is removed by holding the bullion just above the melting point and skimming copper dross from the surface. Agitation, with addition of elemental sulfur, causes the remaining copper to rise to the surface as a black copper sulfide powder, which is skimmed off. After copper drossing, the temperature of the bullion is raised and the bath agitated to induce surface oxidation. The tin, arsenic, and antimony are oxidized, and the oxides (being insoluble in the bath) rise to the surface with some lead oxide, which is skimmed off. The softened bullion usually is desilverized by the Parkes process of stirring metallic zinc into the bullion. Gold and silver, in that order, combine with the zinc and the resultant alloys on cooling rise to the surface and are skimmed off as gold and silver zinc crusts. The zinc remaining in the lead after desilverizing is removed by vacuum distillation or with caustic by the Harris process. If bismuth above acceptable limits is present, the bullion must be refined by the Betts electrolytic process, as at Trail, British Columbia and East Chicago, Indiana or be removed by the Kroll-Betterton process after desilverization.[85]

C. Uses

An unusual combination of physical and chemical properties has given lead a wide range of industrial uses. The most important of these properties are: Softness and extreme workability, high specific gravity, desirable alloying properties, low initial cost and high recoverability, high boiling point and low melting point, good corrosion resistance, and impenetrability by short wave radiation.

The principal uses of lead are for storage batteries, tetraethyl lead, cable covering, paint pigments, building construction, ammunition, and various alloys, chiefly solder, bearing metals, and type metal. Two of the larger uses—paint pigments and tetraethyl fluid—are dissipative. The greater part of the lead used in other forms is recoverable and usually returns to supply as secondary metal.

Lead is widely utilized alloyed with certain metals, principally antimony and tin. The common alloys of lead are roughly classified into bearing alloys, solders, type metals, and low-temperature-melting alloys. Numerous other alloys of lead are extensively employed in industry. The most notable contain varying quantities of antimony and are used for cable sheathing and storage-battery grids. Lead also finds wide usage in machine brasses and bronze.

Large quantities of lead are consumed in the production of litharge, red lead, white lead, lead chromate, and basic lead sulfate. Litharge is used principally in storage batteries in both the positive and negative plates, either alone or mixed with finely divided metallic lead. Smaller quantities are used in the manufacture of ceramics, lead chromate, varnish, insecticides, oil refining, and rubber. Red lead is used chiefly in storage batteries mixed with litharge and metallic lead in the positive plates and in paints. White lead is used almost exclusively as a paint pigment. Basic lead sulfate is used extensively in leaded zinc oxide and as a stabilizer in certain plastics.

An important but small use is for shielding[86] against certain types of dangerous radiation, chief among these being gamma rays. Where space is at a premium and utmost radiation protection is paramount, lead is prescribed. Another advantage is that lead does not become contaminated and may be used continuously without becoming radioactive and emitting its own harmful rays. For this reason it is important that lead for shielding purposes be free of impurities, particularly those that may become radio-active upon exposure to high-energy radiation. The atomic reactors operated by the Atomic Energy Commission employ combinations of concrete, lead, cadmium, and space to protect operating personnel from all types of radiation, including alpha, beta, gamma, and neutron rays. The two last are particularly dangerous because of their ability to ionize matter and hence injure living tissue.

Lead is particularly effective in absorbing gamma rays, whereas cadmium or a hydrogenous material, such as paraffin or water, is used to shield against neutrons; but since gamma rays are emitted when neutrons are absorbed, it is necessary to stop the gamma ray with lead shielding. Lead containers used in handling and shipping radioactive isotopes at the Brookhaven National Laboratories total over 1,000 tons.

Of the chemicals containing lead, tetraethyl lead—a colorless, heavy liquid with formula $Pb(C_2H_5)_4$ and containing 64 percent lead—is the most important. It is the active ingredient of the principal anti-knock compounds added to gasoline to improve their antiknock qualities and efficiency. Premium gasoline for automobiles may contain 2 to 4 cc of tetraethyl lead a gallon (about 3 grams of lead) while most other gasolines contain up to 1.5 cc a gallon. The improved efficiency of gasoline as a result of this additive results in a huge economy in the quantity of fuel used. The Ethyl Corporation recently announced an organomanganese compound which, when used with tetraethyl lead (TEL), makes additional increments to TEL, effective in improving motor efficiency.

Lead has been used for many years to improve the machineability of brass and steel. Lead Industries Association reported at its 1958 annual meeting that 2 percent of lead by weight greatly improves the performance of titanium at high temperatures because of its diffusion and solubility characteristics.

1. Byproducts, Coproducts, and Relationships to Other Commodities

About 60 percent of the U.S. mine production of lead is mined as a coproduct with zinc and from this mixed ore, in addition to the lead and zinc, important quantities of byproduct silver, gold, copper, cadmium, antimony, bismuth, arsenic, and tellurium are recovered. Although most of the zinc and sometimes the copper occurring in complex ores are differentially concentrated, the lead concentrate produced commonly contains zinc and copper, as well as the other byproduct metals listed above. These are recovered in lead smelting-refining to add both to metal supply and to smelter revenue.

2. Substitutes

There are a number of substitutes for lead in many of its uses. In the storage battery industry, the largest consumer of lead, there are some competitive products; but these have different electrical characteristics, and raw materials for their manufacture are not available to replace any sizable quantity of lead in battery usage. The fuel-cell developed by National Carbon Co., Division of Union Carbide Co., and exhibited at the 1958 Brussels World Fair, may in time be substituted for many uses now met by batteries. There are many forms of fuel cells ranging from uranium powered cells to solar cells.

In the cable-covering industry, the third largest user of lead, a substitute has been developed that gives lead considerable competition. This new covering is extruded polyethylene, which is replacing lead telephone cable sheathing for inside use. There are also various combinations of polyethylene with lead and aluminum sheathing for outside and underground use. Aluminum is used in small quantities to sheath telephone and power cable, but economic factors and technical shortcomings must be overcome before the substitution becomes widespread.[87]

The development of substitutes for lead in paint pigments has proceeded rapidly and in many instances lead has been eliminated from paint formulations. Titanium and zinc pigments are the chief competitors of lead in pigments. A substantial quantity of lead in the building industry could be replaced by aluminum, synthetic resins, and other nonmetallic materials. The use of lead in foil manufacture already has been reduced significantly, mainly through replacement by aluminum foil. Plastic films can also be substituted for lead foil.

Several materials may be used as a substitute for lead in caulking. Nonmetallic fibers and plastic compounds are probably the best materials to be substituted for lead in this application, although aluminum wool also may be utilized. The quantity of lead used in type metal could be reduced substantially by substitution of magnesium plates.

3. Secondary Sources and Recovery

The United States has a large reserve of recoverable lead-in-use. The quantity is estimated to be in the order of 5 million tons of recoverable lead, of which about 1 million tons is in lead storage batteries, 3 million tons in power and telephone cable coverings, and over 1 million tons in railway car bearings, lead pipe, sheet lead, and type metal. This in-use reserve provided the secondary industry with scrap.

4. Reserves

Lead-ore reserves are widely distributed on all continents, with important deposits in the United States, Canada, Mexico, Greenland,

Guatemala, Peru, Bolivia, Argentina, and Chile, in North and South America and in many countries of Europe, Africa, and Asia, as well as in Australia. The Western Hemisphere as a whole, however, consumes but little more than 40 percent of world consumption of new lead and has about 38 percent of known reserves. The total world measured and indicated reserve of lead is about 48.8 million tons. Between 70 and 80 percent of the reserve is judged to be recoverable. The reserve of inferred ore is probably as much or more. Of the world's measured and indicated lead reserve, approximately 14.6 million is in North America, 2.5 million in South America, 13.7 million in Europe, 3.5 million in Africa, 2 million in Asia, and 12.5 million in Australia.

North America has about 30 percent of the world's measured and indicated reserve or about 14.6 million tons of lead. Of this quantity, the United States has about 2.9 million tons, Canada about 8 million, and Mexico about 3.5 million. The inferred reserve in the United States is estimated at about 2.7 million tons, while inferred reserves of Canada and Mexico are believed to be at least as large as their measured and indicated reserves. Since 1950 prospecting, exploration, and preliminary development in Canada have indicated approximately 4.4 million tons of lead in lead-zinc ores in the Pine Point area of Northwest Territories and the Bathurst area of New Brunswick.

Peru, Argentina, and Bolivia have the largest lead reserves in South America, but there is considerable evidence to indicate that Chile may have an important, but as yet unrealized, potential. Lack of development in most South American countries suggests that larger reserves will be developed.

Mercury Production and Manufacturing

A. Mercury

Mercury possesses the distinction of being the only metal that is liquid at ordinary temperatures. This property, coupled with mercury's high density, uniform expansion rate, and other useful properties, has enhanced the metal's importance for more than 20 centuries.

The principal commercial mineral of mercury is the red sulfide, cinnabar, that usually occurs in small irregular deposits at comparatively shallow depths. Although cinnabar occurrences are known throughout the world, those of economic importance are limited to a few areas. Mercury production is relatively small and in quantity ranks tenth in world output of nonferrous metals. Italy and Spain dominate world mercury production. Other countries that produced smaller but significant quantities included China, Japan, Mexico, the Philippines, United States, U.S.S.R., and Yugoslavia.

Consumption of mercury tends to fluctuate owing to erratic demands. The high continuing mercury consumption is due mainly to the large quantities used for electrolytic preparation of chlorine and caustic soda, electrical apparatus, industrial and control instruments, in chemicals for agriculture and industrial purposes. Appreciable but smaller quantities are used in precision castings, dental preparations, mercury boilers, and general laboratory applications.

For many of mercury's uses, satisfactory substitutes are known or could be developed; however, for those uses that depend upon mercury's liquidity at ordinary temperatures, high specific gravity, and electrical conductivity, there are few satisfactory substitutes. Consequently, large-scale substitution is deterred except when shortages of mercury are severe and continued.

The United States is not self-sufficient in mercury and depends on imports mostly from Italy, Mexico, Spain, and Yugoslavia, for part of the supply. To bolster domestic production, to provide stability to the industry, and to assure an adequate supply of mercury, the government has programs providing exploration assistance, a guaranteed purchase price, and stockpiling.

Production and consumption of mercury are expected to rise both in the United States and in foreign countries. However, the increases will not be abrupt unless new uses requiring large quantities of mercury are widely adopted. Also, unless more efficient methods are developed for processing the ore, mercury prices will probably rise. The domestic mercury industry is confronted with several problems: principally the necessity of finding additional ore reserves, impediments to efficient and low-cost procedures for recovering mercury, and the lack of a stable market.

B. Background
1. Size and Organization of the Industry
World production of mercury is relatively small and ranks tenth in quantity of output among the nonferrous metals. In the United States production of mercury at individual mines ranges from a few to about 8,000 flasks, and usage at consumers ranges from a few pounds to about 10,000 flasks a year. Only one of the domestic mercury producers is considered a vertically integrated company. The other producers sell virgin mercury through brokers and dealers or directly to consumers. Some consumers act also as brokers and dealers.

Foreign mercury operations differ from those in the United States, principally in the number, size, and ownership of mines. In most instances, only one or two mines furnish virtually the entire output of a foreign country; in addition, the mines in Italy, Spain, and Yugoslavia are wholly or partly state-owned and controlled.

2. Cartel
Mercurio Europeo, a cartel of Spanish and Italian producers, was formed in 1928 when world stocks of mercury were excessive. The aims of the organization were distribution of sales, control of production, and stabilization of prices. Spanish and Italian producers controlled over 80 percent of world production at the time; sales were allocated 55 percent to Spain and 45 percent to Italy. The cartel was formally disbanded January 1, 1950, after the U.S. government purchased a large quantity of Italian mercury with counterpart funds.

Although it has been reported from time to time that the cartel was to be reconstituted, these statements have been denied by the principals. Furthermore, in January 1958 it was announced that mercury production of the Monte Amiata and Siele operations in Italy would be marketed by a joint sales office, Mercurio Italiano.

3. Geographic Distribution of the Industry

Mercury occurrences are known in many countries; however, only a few produce mercury, and of these, Italy and Spain dominate world production. Virtually all the Spanish production comes from one large mine, the Almaden, Province of Ciudad Real, and Italian production from two large mines in the Monte Amiata area. Other major producers are China, Mexico, Japan, the United States, U.S.S.R., Yugoslavia, and the Philippines, which became a mercury producer for the first time in 1955.

The principal mercury-producing states are California and Nevada. Smaller quantities of mercury are produced in Alaska, Arizona, Idaho, Oregon, Texas, and Washington. Mercury production has been reported in Arkansas and Utah.[87]

State/County   Mine
Alaska: Red Devil; Schaefer's Cinnabar
California: Kings, Little King (Fredana), Lake Abbott, Napa, Oat Hill San Benito, New Idria, San Carlos, San Luis Obispo, Buena Vista, Santa Barbara, Gibraltar, Santa Clara, New Almaden Mine and dumps; Guadalupe, Sonoma, Mt. Jackson (incl. Great Eastern), Trinity, Altoona
Idaho: Valley, Hermes, Washington, Idaho-Almaden
Nevada: Esmeralda, B & B, Humboldt Cordero; Red Ore, Pershing,   Hillside
Oregon: Douglas, Bonanza, Jefferson, Horse Heaven, Malheur, Bretz

Of the 300 companies which consume virgin mercury, 80 account for 96 percent of the total domestic consumption and are mainly in the eastern part of the United States. About 400 companies throughout the United States also use approximately 20 percent of the total annual consumption in redistilled form.

4. Definition of Terms, Grades, and Specifications

Mercury produced by the mines is known as prime virgin mercury and contains, in most instances, more than 99.9 percent mercury. Virgin metal with a clean and bright appearance contains less than one part per million of any base metal and is satisfactory for nearly all of the uses.

Flasks for prime virgin mercury are made of wrought iron and spun steel. There are many variations in diameter, height, and weight, ranging as follows: diameter from 4 to 7 inches (usually in the upper part of this range); average height, about 12 inches; weight, from 7 to 15 pounds, averaging about 8 pounds.

Reclaimed mercury, generally as clean as the virgin metal, may contain more gold and silver when not subjected to distillation in the reclaiming process. Grades of mercury other than virgin are produced by concerns that reduce the impurities in virgin metal by multiple distillation or other means

C. Technology

1. Geology

The principal ore mineral of mercury is the red sulfide, cinnabar (HgS—86.2 percent mercury and 13.8 percent sulfur). The native metal occurs in some ores; mercury has also been obtained from livingstonite, metacinnabarite, and in approximately 25 other minerals. Pyrite and marcasite and small quantities of other sulfides, such as arsenic and antimony, often are associated with the cinnabar.

Mercury deposits occur in rocks of all ages and kinds, usually at relatively shallow depths, but almost entirely in regions of volcanic activity. The ore bodies are not true veins and were deposited in irregular chambered veins and brecciated zones, as stockworks of minute seams, and as disseminations and replacements in more or less porous rock. The common gangue rocks are limestone, calcareous shales, sandstone, serpentine, chert, andesite, basalt, and rhyolite.

Cinnabar ore may be classified into two general types— disseminated ore, in which the cinnabar has impregnated a more or less fine-grained or highly brecciated gangue, and ores deposited in fissures and cracks of the country rock. The second type of ore merges into the first as the fissures and cracks become very minute. At the other extreme are veins and bodies of almost pure cinnabar sharply separated from the gangue. The porosity of the rock determines the grade of the ore formed; open-textured sandstone or coarse breccia yield generous space for deposition of high-grade ore bodies, whereas receptacle rocks such as shales or schists give little room for deposition. Proper trap structures, such as clay gangue or relatively impervious rock, are necessary for localizing the ore-bearing solutions in the receptacle rock.

2. Mining

Mercury ore is mined by both underground and surface methods. Of the two types, underground mining furnishes approximately 60 percent of the ore and 70 percent of the mercury production in the United States. At the larger underground mercury mines, square-set stoping or some modification is most frequently used, although in some instances, shrinkage and sublevel stopes are employed. After the ore has been broken by conventional drilling and blasting, it is removed by scrapers, by direct drop to draw points, or by mechanical loaders. For short distances, hand-tramming is used, but locomotives are used for longer hauls. The ore is either trammed or hoisted to the surface.

Open pit or surface mining is done by the usual drilling, blasting, and loading. Power shovels or mechanical loaders, usually less than 1 cubic-yard capacity, are used to load gasoline or diesel-powered dump trucks that haul the ore to the extraction plant.

3. Metallurgy

Milling of mercury ore consists of crushing, sometimes followed by screening. The principal purpose of these operations is to reduce the material to a size required for furnacing; however, mercury minerals often crushing easier and finer than the gangue rock can be upgraded some by screening and rejecting the larger pieces of low-grade or barren material; it can also be upgraded by hand sorting. Gravity concentration of some mercury ore by jigs and tables has been attempted, but with little success due to excessive losses in slimes. Concentration of mercury minerals by flotation is efficient and produces a high-grade concentrate. Despite the necessity of fine-grinding and an ample water supply, the treatment of low-grade ores by flotation before roasting or leaching is advantageous.

Mercury is extracted from ore and concentrate by heating in retorts or furnaces to liberate the metal as a vapor, followed by cooling of the vapor and collection of the condensed mercury. Retorts are inexpensive installations for small operations and require only simple firing and condensing equipment. They are best adapted to operations treating 500 pounds to 5 tons a day of highgrade sorted ore. One of the most objectionable and costly features of retorts is the manual charging and removal of material.

For larger operations, either rotary or multiple hearth furnaces with mechanical feeding and discharging devices are preferred. It is unnecessary to size the feed, which may range up to 3 inches for rotary and 1 1/4 inches for multiple-hearth furnaces. Standard furnace capacities vary from 10 to 100 tons a day with larger sizes to suit requirements. Mercury-laden gases pass from the furnaces through dust collectors into the condensers, where the vapor is cooled and the mercury collected. Final traces of mercury in the gases from the condensers are removed in washers, and the stripped gases are discharged through a stack into the atmosphere. The mercury from the condenser always contains some dirt and soot, which are separated by hoeing the mixture with lime. Mercury from this operation is clean and sufficiently pure for marketing.

Mercury also can be leached from its ores and concentrates with a solution of sodium sulfide and sodium hydroxide and recovered as the metal by precipitation with aluminum or electrolysis. Leaching of mercury ores has not been practiced extensively because of reagent-consuming constituents in some ores, irregularity in compositions of ores, and the cost of find grinding. However, recent studies have indicated that some of these objections can be overcome by concentration of the ore by flotation and leaching the resulting concentrate.

D. Mercury Poisoning

Poisoning may occur in mining and extracting metallic mercur, and in any industry in which mercury is used, as a result of handling or exposure to its vapor.

The Bureau of Mines and many other organizations have studied the subject and issued reports. Precautions that should be taken to insure the safe mining and handling of mercury include proper ventilation of working areas, use of respirators or gas masks, personal cleanliness, spraying contaminated surfaces with various compounds to render the mercury inert, use of mercury vapor detectors, and periodic physical and dental examinations.

E. Uses

One of the larger and growing uses of mercury is as a cathode in the electrolytic preparation of chlorine and caustic soda. Actual consumption of mercury in this manufacturing process is small, although large quantities are required for the original installation.

Large quantities of mercury are used also in electrical apparatus and in industrial and control instruments. Applications include use of mercury in various kinds of lamps, arc rectifiers, batteries, switches, thermometers, barometers, and related equipment. Significant quantities of mercury are used in dental preparations, mercury boilers, as catalysts, and in general laboratory applications; small quantities are consumed in amalgamation and manufacturing fulminate for detonators. A relatively new application of mercury is in frozen mercury patterns for precision casting. Owing to its smooth surface and low uniform expansion upon heating, mercury is superior to wax or plastic pattern materials.

Compounds of mercury are used extensively in insecticides, fungicides, and bactericides for agricultural and industrial purposes. Pharmaceuticals, the control of slime in the pulp and paper mills, and protective paints are smaller, but appreciable quantities of mercury compounds.

Uses that may become important are catalysts in manufacturing organic compounds, pigments, and new applications of amalgam metallurgy. The first two uses were announced recently and have not been evaluated. The latter use will depend upon results of current and future research studies.

1. Substitutes

Other processes are available to replace those using mercury as a cathode in the electrolytic preparation of chlorine and caustic soda and in boilers for generating power. For agricultural and industrial purposes, copper and numerous organic chemicals may be substituted for mercury compounds. In the pharmaceutical field, sulfa drugs, iodine, and various antiseptics and disinfectants may be used instead of mercury chemicals. Plastic and copper oxide paints may be used instead of mercury paint to protect ship bottoms. In some dental uses, metal powders, porcelain, and plastic materials are preferred to mercury amalgams. Lead azide, diazodinitrophenol, and other organic initiators may replace mercury fulminate.

For those applications in electrical apparatus and industrial and control instruments which depend on mercury's high specific gravity, fluidity at normal temperature, or electrical conductivity, satisfactory substitutes are few.

2. Secondary Mercury

Output of secondary mercury furnishes a significant part of the domestic supply. Virtually all metal can be reclaimed from mercury cells, mercury boilers, electrical apparatus, and control instruments, when the plant or equipment is dismantled or scrapped. Other important sources of secondary metal include dental amalgams, batteries, and sludges from electrolytic processes that use mercury as a catalyst.[87]

Cadmium Production and Manufacturing

A. Cadmium

Cadmium rods in the first uranium pile held control of the power that was released from uranium nuclei. A comparative newcomer to the metals family, cadmium thus attained a new significance among metals useful to mankind.

The major uses of cadmium have remained virtually unchanged since the metal was first produced in the United States in 1907. Cadmium protects steel from corrosion, finds use in low-melting alloys and bearing alloys, and can form sulfide compounds, which are the important pigment materials in yellow and red paints, printing inks, paper, soap, textiles, and other products. Three new, unusual uses have appeared in recent years, which may prove to be large-scale users of cadmium:

o Cadmium sulfide converts light energy directly into electrical energy; it may be the harness for solar energy.

o Cadmium sulfide phosphor, upon absorption of beta radiation, emits light which can be picked up by a photoelectric cell and translated into electricity.

o Cadmium metal readily absorbs thermal neutrons, which permits it to be used to control and shield atomic reactors.

Cadmium is a soft, silver-white metal. It exists in the earth's crust in com-pound form to the extent of about 0.15 parts per million. No ore is mined solely as a source of cadmium. Commercial production is derived from greenockite, the sulfide of cadmium, which is found sparsely associated with zinc ore, or with other ores that contain some zinc, such as those of copper and lead. The cadmium content of zinc concentrates normally ranges from 0.17 percent to 1.4 percent.

The United States is the world's leading producer of cadmium. From 1940 domestic plants produced approximately two-thirds of the total world output. Since 1940 about 60 percent of the primary cadmium metal produced in the United States has been derived from materials of foreign origin. Mexico is the chief foreign source, with Canada, Australia, Belgium-Luxemburg, Italy, and Southwest Africa supplying large quantities. Eleven zinc smelting companies produced primary metallic cadmium at 14 plants in the United States.

Cadmium problems include: dependence of production on output of zinc, undelineated loss of cadmium in ore dressings, limited technology in secondary recovery, restricted use owing to the high price of cadmium compared to that of competitive metals, and inadequate statistics for distribution and consumption.

B. Background[88]

1. Size and Organization of the Industry

The production segment of the U.S. cadmium industry has expanded with growing consumer demand over the years since 1907. In that year production totaled 14,000 pounds of the metal compared with about 10 million pounds produced in 1956. Domestic plants have turned out nearly 198 million pounds of primary during the past 51 years. The cadmium industry of the United States includes primary and secondary metal producers, distributors, and consumers. Primary metal production is of pre-dominant importance, supplying over 95 percent of the total annual output. Relatively small quantities of cadmium compounds are produced from primary raw materials; such production is in addition to the new metallic supply. Recovery of cadmium metal and compounds from scrap materials constitutes a comparatively minor segment of the cadmium industry today. Cadmium distributors are divided into two general groups: (1) the companies that, for the most part, produce plating salts, supply special platers' shapes as well as regular grades of cadmium and provide technical service for their customers who are chiefly electroplaters and (2) buyers of domestic and imported cadmium for resale. There are thousands of consumers of cadmium, most of whom use relatively small quantities of the metal. The largest group of consumers use the metal for electroplating various manufactured products. Bearing manufacturers, pigment producers, and solder-makers are the older principle consumers.

Although the cadmium industry of other countries is smaller than that of the United States, it is organized along the same lines. Southwest Africa and Mexico are exceptions, in that cadmium is not refined in these countries.

2. Geographic Distribution of the Industry

Twenty countries of the world produce cadmium. Southwest Africa ranks among the largest cadmium producers by virtue of high cadmium content of the cadmium-bearing ores mined. That country does not produce refined metal; lead-zinc-copper concentrates containing large quantities of cadmium are exported principally to the United States, the United Kingdom, Belgium, and France, where the metals, including cadmium, are recovered.

Mexico, Canada, and Australia, all large zinc producers, are also important cadmium producers. Virtually all Mexican cadmium is exported in fine dust and zinc concentrate, chiefly to the United States. Canada and Australia produce refined cadmium metal; Belgium and the United Kingdom, using imported source materials, are also significant cadmium producers. Japan, Norway, Italy, processing domestic materials, and France treating imports, produce sizable quantities of cadmium. Germany formerly recovered large quantities of cadmium from the cadmium-rich zinc ore of Upper Silesia, but with the cession of that territory to Poland after World War II, Germany has been less important as a source of cadmium.

Approximately 60 percent of the cadmium consuming industry of the United States is in states on the Great Lakes: Illinois accounted for 13 percent, New York 11 percent, Ohio 10 percent, Michigan 8 percent, Pennsylvania 8 percent, Indiana 4 percent, Wisconsin 3 percent, and Minnesota 1 percent; New England contains about 14 percent of the consuming industry, Massachusetts 7 percent, Connecticut 6 percent, Rhode Island 0.6 percent, and New Hampshire 0.4 percent. The remaining 28 percent of the consuming industry is distributed among 37 states, including the District of Columbia.

3. Definition of Terms and Grades

Primary cadmium metal and cadmium compounds are produced directly from cadmium-bearing zinc concentrate, flue dust, and other primary material. Secondary cadmium metal is recovered from old cadmium-base bearings, cadmium-bearing alloys, skimmings, and drosses. Secondary cadmium compounds are those derived from scrap or secondary cadmium and include compounds such as cadmium nitrate, sulfide, oxide, hydroxide, chloride, and carbonate.

Electrolytic cadmium is 99.95 percent pure. Cadmium anodes in special platers' shapes are those that have been given a particular form, such as a bar of elliptical cross section, tuning fork, or hexagonal bar.[89] The object of giving the anodes particular shapes is to increase the available surface area exposed to the solution for a given weight of anode in order to increase the economy of the plating operation.

C. Technology

1. Geology

Cadmium sulfide is found as a crystalline (greenockite) or an amorphous (xanthochroite) deposit on sphalerite, or in solid solution with sphalerite. Weathering causes cadmium sulfide to oxidize into cadmium oxide or cadmium carbonate (otavite). The earth's crust is said to contain 0.15 parts per million of cadmium. Zinc ore, when concentrated, has been found to contain from trace quantities to about

1.4 percent cadmium. No deposit has been found in which cadmium was the principal valuable constituent.

2. Recovery Process

As the primary supply of cadmium is derived mainly from zinc ores, production of cadmium metal is intimately tied to the processing of such ores. Basically, the process for recovering cadmium from zinc ores and concentrates comprises the following steps: (1) roasting the ore or concentrate; (2) centering the resulting calcine and dust, along with cadmium furnace residues, a mixture of sodium chloride or zinc chloride and zinc sulfate liquor, and fuel; (3) sulfuric acid leaching of the cadmium-rich Cottrell dust from the centering machines; (4) precipitating the cadmium as a sponge from the zinc-cadmium solution by adding zinc dust; and (5) distillation of the sponge in a horizontal, batch-type retort, yielding cadmium metal, and residue. At electrolytic zinc plants recovery of cadmium is briefly as follows: the zinc and cadmium are leached from the roasted concentrate and flue dusts to produce a zinc electrolyte. Before electrolysis the solutions are purified, the cadmium being precipitated with zinc dust and filtered from solution for subsequent recovery by distillation or electrolysis.

An important source of cadmium is the baghouse dust from copper and lead smelters operating on zinc-bearing ores containing cadmium. Recovery is effected by several methods. In one process the fired baghouse dust of the lead blast furnace is re-fumed, and a concentrated cadmium dust is recovered and sent to an electrolytic plant for further treatment. In another method baghouse dust is concentrated by fuming off the cadmium compounds (oxide, sulfate, sulfide, chloride) in a reverberatory furnace. The concentrated fume is acid-leached, and the cadmium is precipitated from the pregnant leach liquor with zinc. The sponge metal so obtained is distilled to recover metallic cadmium. By another process, which is applicable to both zinc and lead-smelter products, pulped baghouse fume is treated with chlorine gas, and solubilized cadmium is precipitated by zinc dust. The precipitated cadmium sponge may be separated by filtering or centrifuging and electrolyzed after solution in sulfuric acid, or it may be dried in the absence of free oxygen and distilled.

New technology being developed in research laboratories is designed to increase recovery of cadmium from ore milling slimes. Essentially this work involves separation of the sulfides of cadmium, lead, indium, and germanium from sphalerite by selective distillation in an inert atmosphere. It is believed that success of this research would add substantially to the supply of cadmium.

D. Uses

The uses of cadmium divide into four general categories: electroplating, pigments and chemicals, alloys, and storage batteries. Cadmium used in electroplating is distributed among fasteners of various kinds, machinery parts, communication and electrical equipment, building materials, and transportation equipment. Cadmium metal plating protects the structural material from corrosion. Cadmium pigments and chemicals are used as protective and decorative coatings on metal and wood; to color plastics; as an aid in photography, process engraving, and lithographing; and to convert energy in the form of light to electricity. Cadmium alloys with various metals such as tin, lead, copper, magnesium, nickel, silver, zinc, indium, and mercury for use in such things as bearings, low melting alloys, amalgams, and electricity conductors. Alkaline storage batteries use cadmium in the negative plate of the cells. Small quantities of cadmium metal and cadmium alloy are used in nuclear reactors.

1. Substitutes

Zinc and chromium-nickel can substitute for cadmium as a plating material to protect iron and steel from corrosion. Tin and bismuth can be used in place of cadmium in low-melting alloys. The chromates of lead, zinc and barium can, in some instances, replace cadmium yellow as a pigment. In the field of atomic energy cadmium control rods, which limit neutron activity in an atomic pile, can be replaced by the more expensive boron, hafnium, samarium, europium, and gadolinium; as a neutron shield, boron or hafnium can be used in place of cadmium.

2. Reserves

There is no reserve of cadmium ore as cadmium is recovered only as a by-product of zinc. The cadmium content of measured and indicated world zinc reserves is thought to be about 811 million pounds. It is estimated that Canada has 68 million pounds, Australia 50 million pounds, the United States 67 million pounds, Mexico about 97 million pounds, other free world countries a total of 699 million pounds, and Russia 112 million pounds.

Zinc concentrate from the tri-state area of Missouri, Oklahoma and Kansas average about 0.35 percent cadmium; mines in the western states rarely carry more than 0.25 percent. Zinc concentrate from east Tennessee ores contain about 0.4 percent cadmium. Zinc concentrates from Canada contain an average of 0.25 percent cadmium, Australia 0.20 percent, Mexico 0.55 percent, Bolivia 0.27 percent and those from Peru range from 0.17 to 1.4 percent. A minor cadmium resource exists in cadmium-bearing scrap materials such as old bearings, low melting point alloys, and secondary residues.

E. Production, Consumption, and Foreign Trade

The United States is the world's largest producer of cadmium. Automotive parts, communication devices, aircraft parts, electrical equipment, and hardware account for 75 percent of cadmium used. The remaining 25 percent is used to plate parts used in building materials, ordnance, office machinery, textile machinery, household appliances, ships, control instruments, bicycles, heating and refrigeration equipment, amusement and vending machines, industrial machinery, petroleum industry equipment and various kinds of medical, health and safety equipment.

Approximately 7 million pounds of cadmium are consumed annually in dissipative uses including protective plating and in pigments. Airplanes and refrigerated railroad cars in growing numbers use nickel-cadmium batteries for power supply. These batteries require about 0.008 pounds of cadmium per amperehour per cell. Roughly 60 percent of the primary metallic cadmium produced in the United States is from foreign materials, although a lack of data precludes determination of the cadmium recovered from imported materials by country. The cadmium content of imported flue dust is recorded, but the weight of the flue dust is not. The approximate cadmium content is available for imported zinc ores and concentrates. Almost all the imported flue dust-bearing cadmium and a large proportion of the zinc ores and concentrates come from Mexico. Canada, Peru, and Southwest Africa are other important sources of cadmium-bearing raw materials.

Technical Terms, Chapter III

amalgan: an alloy of mercury and other metals

amorphous: without shape or form

amyotrophic lateral sclerosis: degenerative disease of the motor neurons

analagous: similar to, or like

angina pectoris: a paroxymal thoracic pain

asymptomatic: without symptoms

barite: native barium sulfite

bismuth: metallic element

calcine: to reduce to powder

carcinogenic: ability to cause development of cancer

cardiomyopathy: any disease that affects the structure and function of the heart

chelation: treatment for heavy metal poisoning

cytotoxic: agent that destroys or damages tissue or cells

deleterious: to hurt or damage

dioxins: a contaminent of a herbacide

etiology: study of factors involved in development of a disease

europium: metallic element of the rare earth group

extrapolation: to project by inference (a theory)

gadolinium: a rare earth metalic element
greenockite: the sulphide of cadmium
hafnium: quadrivalent metal closely resembling zirconium
idiopathic: without a known cause
imperil: to bring into peril, to endanger
litharge: a fused form of lead monoxide
livingstonite: a lead, gray mercury antimony
marcasite crystallized iron pyrites
metacinnabarite: native mercuric sulphide
metallurgy: theoretic and applied sciences of the nature and uses of metals
methylmercury: gaseous mercury compound
microammeter: an instrument for measuring microamperes
oral galvanism: electricity produced in the mouth by dissimilar metal and acid or aklaline saliva
phlebitis: inflammation of a vein
plethora: fullness
polyneuropathy: a condition in which many nerves are afflicted with a disorder
porphyria: genetic blood disease
pseudoscientist: self-styled scientist
pyrite: a common mineral, metallic finish
rhonochrosite: mineral compound
samarium: pale grey, rare metallic element
sapa: grape juice evaporated to a syrupy consistency
scanning electron microscopy (SEM): technique using a scanning electron microscope on an electrically conducting sample
sidrite: mineral compound
silicates: glass-like material
sphalerite: ore of zinc
thioethers: a sulphur ether
thorium: readioactive metallic element
tungsten: a metallic element

References, Chapter III

Adam, R. D. and Victor, M. *Principles of Neurology, Fifth Edition*. 1993, p. 778.

Amin-Zaki, L. et al., "Intra-Uterine Methyl Mercury Poisoning in Iraq," Pediatrics, 54 (5), November 1974, p. 587-595.

Brune, C. et al., "Gastrointestinal and In Vitro Release of Copper, Cadmium, Indium, and Zinc from Conventional and Copper-Rich Amalgam." Scand. J. Dent. Res., 1983, 91: pp. 66-71.

Choi, Ben H., "Methylmercury Poisoning of the Developing Nervous System: Pattern of Neuronal Migration in the Cerebral Cortex," Neurotoxicology, 7(2): 591-600 (1986).

Corbin, S. B. and Kohn, W. G., "The Benefits and Risks of Dental Amalgam: Current Findings Reviewed," Journal of the American Dental Association, Vol. 125, April 1994, p. 383.

DHEW Publ. No. (NIH) 74-473, 1974."Cardiotoxic Effects of Mercury," Ch. XI, pp. 199-200.

Eggleston, D. W. and Nylander, M., "Correlation of Dental Amalgam with Mercury in Brain Tissue," Journal of Prosthetic Dentistry, Vol. 58, No. 6, p. 704-707, December 1987.

Environmental Law Foundation, News Release, December 14, 1993,"Proposition 65 Warnings for Mercury in Dental Fillings," p. 1-2.

Eyl, Thomas B., "Methyl Mercury Poisoning in Fish and Human Beings," Modern Medicine, November 16, 1970, p. 135-141.

Gay, Cox, and Reinhard, "Chewing Releases Mercury from Fillings," Lancet, 985, May 5, 1979.

Huggins, H. A.(1993). *It's All in Your Head: The Link Between Mercury Amalgams and Illness*. Paragon Press, pp. 41-45; 78-80

Huggins, H. A.(1989). *It's All in Your Head: Diseases Caused by Silver-Mercury Fillings*. Life Sciences Press, p. 28.

Huggins, H. A., *The Applications Textbook*. (1988) p. 365; pp 372-373

Huggins, H. A., unpublished data.

Kasper, E. K. et al., "The Causes of Dilated Cardio-myopathy: A Clinico-pathologic Review of 673 Consecutive Patients." JACC, Vol. 23, No. 3, March 1, 1994: 586-590.

Kuntz, W. D. et al., "Maternal and Cord Blood Background Mercury Levels: A Longitudinal Surveillance," Am. Jour. Obst. and Gynec., June 15, 1982, 143: 440-443.

Marlowe, M. et al., "Main and Interaction Effects of Metal Pollutants in Emotionally Disturbed Children," Monograph in Behavioral Disorders, R. B. Rutherford (Ed.), (1984), Vol. 7, Reston, Virginia: Council for Children with Behavioral Disorders, p. 67-79.

Meinig, G. E., "Root Canal Cover-up Exposed!"(1993). Bion Publishing, Ojai, CA pp. 75; 120-121.

Placidi, G. F. et al.(1983). "Distribution of Inhaled Mercury (203 Hg) in Various Organs." Int. J. Tiss. React., 5: 193-200.

Queen, H. L., Mercury-Free News, Vol. 5, No. 2, May 1992, p. 10.

Queen, H. L.(1988). *Chronic Mercury Toxicity: New HopeAgainst an Endemic Disease.* p. 75. Queen and Company Health Comm, Inc., Colorado Springs, CO.

Queen, H. L., Heart Talk. February 1989: Vol. 8, No. 1, p. 5. San Francisco Examiner, "Dentists to Post Warnings on Mercury," December 15, 1993, p. A-1.

Shiraki, H. & Nagashima, K.(1977) "Essential Neuro-pathology of Alkyl-mercury Intoxications in Humans from the Acute to the Chronic Stage with Special Reference to Experimental Whole Body Autoradiographic Study Using Labeled Mercury Compounds." Neurotox., Ed.: Roizin, L., et al., 247-260, 1977.

Townsend Letter for Doctors (April, 1994). "California Dentists Now Required to Inform Patients of Risks of ToxicSubstances in Dental Fillings." No. 129, p. 394. Port Townsend, WA.

Trakhtenberg, I. M., *Chronic Effects of Mercury on Organisms,* Chap. VI: pp. 109-134, "The Micromercurialism Phenomenon in Mercury Handlers." DHEW Publ. No. (NIH) 74-473, 1974.

Vimy, M. J. and Lorscheider, F. L., "Serial Measurements of Intra-Oral Air Mercury: Estimation of Daily Dose from Dental Amalgams," J. Dent. Res., 64(8): 1072-1075, August 1985.

Virkkunen, M. "Behavioral Changes Relating to Changes in Total Serum Cholesterol," JAMA, 253(5): February 1, 1985.

Weening J. J., Chapter 4: Mercury Induced Immune Complex Glomerulopathy: An Experimental Study, Van Dan Denergen, 1980.

Whitney, H., "The Biological Roller Coaster: Chronobiologists Study the Body's Natural Rhythms." Omni, February 1994: Vol. 16, No. 6, p. 26.

Wyngaarden and Smith, W. B. (Eds) (1985). *Cecil Textbook of Medicine*. Louria, D. B., "Trace Metal Poisoning," p. 2309. Saunders Co.

Ziff, S., "Silver Dental Fillings: The Toxic Time Bomb." Aurora Press, 1984, p.8-9. Huggins Diagnostic Center, Position Papers: "Amalgam Issue, Root Canals, Cavitations." p. 15.

CHAPTER IV
Conclusions, Recommendations, and Implications

Until recently, textbooks and governmental guidelines alike claimed that kidney disease due to occupational lead exposure rarely occurred in the United States because of this country's advanced industrial health practices.[90] It was generally agreed that lead intoxication was seen chiefly in the industrialized nations of Europe where less effective occupational health standards prevailed. Lead nephropathy was relegated to the distant past and to the underdeveloped countries of the world. Kidney damage was believed to occur only when the blood lead was markedly elevated.[91]

A young man had been hospitalized with abdominal pain and anemia. This was his sixth hospital admission for similar complaints over two years. With each hospitalization a different diagnosis had been made: hepatitis, ulcer, gastritis, gall bladder, virus, and neurosis. His gall bladder had been removed with subsequent relief of the pain. In fact, any in-hospital treatment seemed to relieve the excruciating abdominal cramps, but a month or two after returning to work his pains recurred. For five years he had manufactured solder to be used for electrical connections. He prepared solder creams by converting molten lead into powder in a small unventilated room. One physician, suspecting lead poisoning, had drawn blood for measurement of lead, but the patient's blood level was reported to be within what was then an "accept-able" range: less than 80 ug of lead per 100 ml. The diagnosis of lead poisoning could not be established.[98]

Methodology to prove the diagnosis of asymptomatic lead poisoning could not be found in the federal guide-lines but had to be sought elsewhere. To the traditional symptoms of colic followed by the palsy, modern biochemists had added sophisticated laboratory test of blood and urine. Anemia and immature red blood cells had been associated with plumbism in the nineteenth century. The defects in red cell production induced by lead could now be detected by exquisitely sensitive tests of heme synthesis. Highly accurate measurements of the blood lead concentration were also available.

Tests of urine and blood had found favor because of convenience—the accessibility of both methodology and the body fluids. But lead differs from most environmental toxins in that once it enters the body it is stored in the bones. Over 95 percent of the body lead is retained in bone where it persists for many years. The biological half life of bone lead is estimated to be several decades. This lead storage provides the potential for both delayed toxicity and diagnostic assessment.

The standard technique for detecting excessive lead absorption—determination of the blood lead concentration—was grossly inadequate to identify workers at risk for occupational lead nephropathy. The insensitivity of the blood lead to the level of exposure had been one of the better kept secrets of the lead industry for fifty years.[98] Because many of the men studied had terminated their lead exposure months earlier, the EDTA lead mobilization test was required to find those with excessive body lead stores. Over five years, 21 cases of unsuspected renal disease among lead workers were discovered. In 15, no other cause of renal disease could be found, and in four, the lead nephropathy was reversed by long-term chelation therapy.[98]

In 1978, OSHA released new regulations for lead exposure in the workplace.[90] One of the most important contributions to the understanding of adverse health effects associated with exposure to inorganic lead was the evidence on kidney disease during the hearings. It is apparent that kidney disease from exposure to lead is far more prevalent than previously believed. In the past, the number of lead workers with kidney disease in the United States was thought to be negligible, but the record indicates that a substantial number of workers may be afflicted with this disease. A nephrologist (kidney specialist), who testified at the hearings for OSHA, stated that a minimal estimate of the incidence of this disease (nephropathy) would be 10 percent of lead workers. According to this estimate, "there may be 100,000 cases of preventable renal disease in this country... If only 10 percent of these hundred thousand workers came to chronic hemodialysis (kidney machines), the cost to Medicare alone would be about 200 million dollars per year."[98]

The "Final Standard" presented a detailed analysis of the conclusions from preliminary studies. By the time lead nephropathy can be detected by usual clinical procedures, enormous and irreparable damage has been sustained. The lead standard must be directed towards limiting exposure so that occupational lead nephropathy does not occur, since in this situation progression to death or dialysis is likely. The record indicates that blood lead is an inadequate indicator of kidney disease development, since rather than being a complete measure of body burden, it is merely a measure of absorption when sampled close to the time of exposure. Given these conclusions, OSHA must approach the prevention of kidney disease by recognizing the limited usefulness of certain biological parameters.

The lead standard must therefore be directed towards limiting exposure so that occupational lead nephropathy is prevented. Lead nephropathy is important because the worker has lost the functional reserve, the safety provided by two normal kidneys. If one kidney becomes damaged, the normal person has another to rely upon. The lead worker with 50 percent loss of kidney function has no such security. Future loss of kidney function will normally occur with increasing age, and may be accelerated by hypertension or infection. The usual life processes will bring the lead worker to the point of uremia, while the normal individual still has considerable renal functional reserve. Loss of a kidney is therefore more serious than loss of an arm, for example. Loss of an arm leads to obvious limitations in activity. Loss of a kidney or an equivalent loss of kidney function means the lead worker's ability to survive the biological events of life is severely reduced.

In preparing the "Final Standard" a new philosophy was formulated concerning the responsibility of government in protecting the health of industrial workers. Rather than awaiting the onset of clinical symptoms, OSHA had accepted responsibility for preventing organ damage. The crucial concept is that subtle, but measurable, biologic derangements are considered evidence of "materially impaired health," even when detected only in the laboratory. The worker, then, does not have to be symptomatic to be recognized as ill. Occupational health specialists are more accustomed to cancer incidence data than to physiologic estimates of kidney function, but evidence that lead causes renal cancer in man is virtually nonexistent.[92] The regulatory bodies were therefore forced to come to terms with the more abstract measurements of pre-clinical organ dysfunction in the blood, nervous system, and the kidneys. Impairment of the enzymes responsible for synthesis of hemoglobin has been demonstrated at blood lead levels as low as 18 ug/dl.[93] Public pressure to reduce the "acceptable" blood level towards 18 ug/dl may soon prove irresistible. The potential financial impact of the federal regulations did not escape the attention of industry. The lead companies sued. On June 29, 1981, the United States Supreme Court denied the industry's claims; the Court upheld the obligation of OSHA to protect industrial workers without first providing proof of "cost-effectiveness." Defeat in court shifted the lead industry's attacks directly to the regulatory agencies themselves.

With the publication of the "Final Standard," American medical opinion on lead nephropathy reverted to that of the nineteenth century. Occupational lead nephropathy once again received official recognition.[94] Such turnabouts do not come easily but are by no means unheard of in medicine. Unconventional ideas have never been eagerly received by physicians. Orthodoxy is impressed upon the biomedical community by the presence of "experts" on research grant and editorial review panels. Reviewers resist new ideas for three reasons:

(1) If the expert agrees, he would have performed the studies himself.

(2) If he disagrees, his own theories are, in all likelihood, in jeopardy.

(3) Defects tend to be more glaring to the reviewer in work which refutes his own point of view than in work which supports him.

Expertise itself introduces conservatism, if not bias. An old boy network exists wherein reviewers are more likely to trust the assertions of investigators with whom they share academic experience and insight. The proponents of conventional work are supported, while delays in funding or publication of unorthodox ideas may cause unconventional research to perish. In the field of environmental toxicology, unadvertised financial support of selected experts by the lead industry, both directly and indirectly, makes it particularly difficult to gain acceptance of new ideas which might increase manufacturing costs.

Conventionality in research, however, is not entirely useless. Maverick ideas can only be incorporated into a structure of existing knowledge. Deviations from the norm are properly viewed with suspicion. Science requires consistency: a cautious and conservative safeguard against blunder. New concepts, like radical politics, meet resistance until internalized by the establishment.

The major difference in the adoption of unconventional ideas in the twentieth, as compared to the sixteenth century, is the speed of change. Shifts in thinking occur in decades rather than centuries, as conceptualization forever chases technology. Methodologic advances sire new experimental generations about every ten years: roughly the span required for the limitations of current techniques to be appreciated. Transitions in thinking are rarely abrupt. Conformity is quickly reestablished, as last year's renegade hypothesis is adapted and adopted, fashioned into new conventionality. Conceptual evolution has been telescoped from centuries to decades, but experimental surprises remain inherently unwelcome. Resistance to change is easily masked by scientific skepticism, which may serve as a convenient guise for obstructionism.

Knowledge of the impact of lead on health has evolved over two thousand years and medical resistance to changing concepts of its toxicity has been as constant as the disease itself. Lulled by familiarity, physicians and the public alike are not easily aroused by claims of new dangers from this venerable poison. Vested interests are superimposed on medical complacency. The enormous value of lead to society adds another dimension to the usual resistance to change. Economic and professional bias have bolstered conservative views throughout history while fear of the unfamiliar excites far more attention than fear of the commonplace. But, like science itself, human self-interest shows signs of being self-correcting. The history of industrial health practices suggests that protection of man from man-made hazards is not only possible, but, in the long run, likely to prevail.

A. Chemical Injury
　1. Mechanisms of Chemical Injury
　Chemical injury begins with a biochemical interaction between a toxic substance and the cell's plasma membrane, which is ultimately damaged, leading to increased permeability. Not all of the mechanisms causing chemically induced membrane destruction are known.

　2. Chemical Agents
　A number of chemical agents cause cellular injury. Highly toxic substances are known as poisons. Minute amounts of some, such as arsenic and cyanide, can rapidly destroy enough cells to cause somatic death (death of the individual). Chronic exposure to air pollutants,

insecticides, and herbicides can cause cellular injury. Carbon monoxide, carbon tetrachloride, and social drugs, such as alcohol, can significantly alter cellular function and injure cellular structures. Over-the-counter and prescribed drugs may also cause cellular injury, sometimes leading to death. In addition, accidental or suicidal poisonings by chemical agents cause numerous deaths. The injurious effects of some of these agents—lead, carbon monoxide, hydrogen cyanide, and hydrogen sulfide—exemplify common cellular injuries.

3. Lead

Heavy metals, such as lead, cause a significant number of childhood poisonings. Lead-based paint, which has a sweet taste, is often ingested by children. Children are particularly vulnerable to lead toxicity because, compared to adults, they absorb lead more readily through the intestines.[99] If nutrition is compromised, especially if dietary intake of iron, calcium, and vitamin D is insufficient, lead's toxic effects are enhanced.

The organ systems primarily affected by lead include the nervous system, the hematopoietic system (tissues that produce blood cells), and the kidneys. A suggested mechanism by which lead acts on the central nervous system is interference with neuro-transmitters, which may cause hyperactive behavior.[100] Lead inhibits several enzymes involved in red blood cell synthesis.[101] A significant manifestation of lead toxicity is anemia caused by lysis of red blood cells (hemolysis). Manifestations of brain involvement include convulsions and delirium and, with peripheral nerve involvement, wrist, finger, and sometimes foot paralysis. Renal lesions can cause tubular dysfunction resulting in glycosuria (glucose in the urine), amino-aciduria (amino acids in the urine), and hyperphosphaturia (excess phosphate in the urine). Gastrointestinal symptoms are less severe and include nausea, loss of appetite, weight loss, and abdominal cramping.

4. Ionizing Radiation

Ionizing radiation is any form of radiation capable of removing orbital electrons from atoms. Ionizing radiation is emitted by x-rays, gamma rays, alpha and beta particles (which are emitted from atomic nuclei in the process of radio-active decay), and from neutrons, deuterons, protons, and pions—all of which are emitted from cobalt or linear accelerators. Occupations exposure to ionizing radiation is mostly limited to alpha and beta particle exposure and exposure to x-rays, gamma rays, and neutrons. Radiant energy from sunlight (solar radiation) can also injure cells.

The most abundant source of exposure to ionizing radiation is the environment. This source includes emissions from radioactive material inside the body, cosmic rays from outer space, and radiation emitted from such substances as soil and building materials. Environmental radioactivity is primarily emitted by uranium, thorium, and potassium. Other sources are x-rays used for medical diagnosis and treatment, uranium and thorium mines, nuclear weapons, and nuclear reactors that generate electricity.

The mechanism by which ionizing radiation damages cells is the most vulnerable target of radiation, particularly the bonds within the DNA molecule. Membrane molecules and enzymes are also damaged by radiation. The nuclear changes include nuclear swelling, disappearance of the nuclear membrane, and pyknosis (a cellular degeneration in which the nucleus shrinks and chromosomes deteriorate). The chromosomal aberrations include breaks, deletions, translocations, and many other structural abnormalities. The intensity, duration, and cumulative effects of exposure to ionizing radiation determine the extent of injury.

Not all cells and tissues have the same sensitivity to radiation, although all cells can be affected. Radio-sensitivity depends on rate of mitosis and cellular maturity. Because fetal cells are both immature and undergoing rapid cycling, the fetus is at great risk for injury caused by ionizing radiation. Particularly vulnerable are embryonic germ cells, which are precursors of ova and sperm. Throughout life, cells of the bone marrow, intestinal mucosa, testicular seminiferous epithelium, and ovarian follicles are susceptible to injury because they are always undergoing mitosis, which ensures the presence of vulnerable, immature daughter cells.

The effects of ionizing radiation may be acute or delayed. Acute effects of high doses, such as skin redness, skin damage, or chromosomal aberrations, occur within hours, days, or months. The delayed effects of low doses may not be evident for years. Effects are usually (1) somatic (involving the exposed individual's entire body), for example, leukemia and other cancers; (2) genetic, involving offspring of the exposed individual; or (3) fetal, involving fetuses that are exposed in utero. Cells that are particularly susceptible to damage are those of the gastrointestinal tract, bone marrow, lymph nodes, fetus, and ovarian follicles.[102]

In issuing the new regulation of March 6, 1996, the government took an important step forward in two areas of critical concern to the Clinton administration—protecting the public from such serious health threats as lead and arming the public with information they need to protect themselves. The EPA recognizes that the public has a right to know about toxic hazards in the community and has expanded that right under the new regulation.[95]

Over the past generation, protecting children from lead poisoning has been one of the great success stories of this nation's commitment to public health and environmental protection. By banning leaded gasoline and lead-based paint used in homes, we reduced the level of lead in the air by 90 percent and protected millions of children from permanent mental and physical harm.[95]

Lead exposure is the biggest environmental risk to children today. Some 1.7 million children still have blood-lead levels that are too high to be safe, according to the U.S. Department of Health and Human Services. Lead poisoning lowers intelligence and causes behavioral problems. And when pregnant women are exposed, lead can cause abnormal fetal development.

All parents have a right to know that where they raise their children is safe. The new regulation will help to ensure that no child will experience the tragedy of lead poisoning ever again. It gives the public the right to know about lead paint hazards when they buy or rent a house—and gives them the common sense information they need to reduce those hazards.[95]

Protecting children from lead poisoning has been one of the great success stories of this nation's commitment to public health and environmental protection. Since the 1970's, federal and state efforts to address dangerous sources of lead have resulted in a 90 percent reduction in the average blood-lead levels in children. Much of this success is the result of the phase out of leaded gasoline, the ban on lead in food cans, and the ban on lead-based paints in residences, EPA researchers have focused their efforts on the young, having long noted that such environmental pollutants as lead can harm the sensitive brain and nervous systems of growing children more severely than adults.

Despite recent success in prevention of lead poisoning, improperly managed lead-based paint in older homes remains the greatest source of exposure for the nation's children. In fact, the Secretary of the Department of Health and Human Services recently characterized lead poisoning as the "number one" environmental threat to the health of children in the United States.[96] In 1992, a survey estimated that more than 3 million tons of lead in the form of lead-based paints still lurked in American dwellings built before 1980.

The Department of Health and Human Services identified some previously unrecognized dangers of lead poisoning and blood-lead levels once considered safe have been called into question. Over the period 1978 to 1991, the then Centers for Disease Control lowered the blood-lead level of concern from 60 to 10 micrograms per deciliter (a millionth of a gram in a tenth of a liter). But even low blood-lead levels have been associated with learning disabilities, growth, hearing and visual impairments, and other damage to the brain and nervous system.

Today, more than 1.7 million American children under age six have blood-lead levels that exceed 10 micrograms per deciliter. Most of those children are being poisoned by deteriorating lead-based paint and the contaminated soil and dust it generates. Lead from exterior house paint can flake off or leach into the soil around the outside of the house. Dust created during normal paint wear (especially around windows and doors) can create a hard-to-see film on house-hold surfaces. Cleaning and renovation can increase the threat of lead-based paint exposure by dispersing fine lead dust into the air.

If lead paint removal is managed improperly, both adults and children can inhale the fine dust or ingest dangerous amounts of paint dust via hands and food. Children under age six are particularly susceptible. But much of the lead in homes can be handled and maintained safely by using simple, low-cost, commonsense procedures.

Recognizing the need for more public education, the Environmental Protection Agency and the Department of Housing and Urban Development jointly issued a regulation in March 1996 providing for disclosure of possible hazards in lead-based paint at the time when homes are sold or rented. The rule implements section 1018 of the Residential Lead-Based Paint Hazard Reduction Act of 1992, designed to protect families from exposure to lead from paint, dust and soil.

This rule represents the federal government's first step under the law of 1992 to build an infrastructure to ensure the elimination of lead-based paint hazards in housing. Other steps, to be implemented later, include oversight of state and local programs for training and certification of contractors who perform lead inspection and abatement services, and working with lending and insurance institutions to make lead safe housing available. The EPA will cooperate with other federal, state, and local government agencies, as well as private organizations in carrying out these additional steps to prevent lead poisoning.[96]

The rule requires that, beginning with the fall of 1996, sellers, landlords, and their agents must disclose information on lead-based paint and lead-based paint hazards to tenants and renters before selling or leasing a home. This information is to include reports that are available from tests that may have been performed before sale or lease, as well as a federal pamphlet with practical, low-cost tips on identifying and controlling lead-based paint hazards, "Protect Your Family from Lead in Your Home." Home buyers will also have a ten-day period to conduct a lead-based paint inspection or risk assessment at their own expense.

As a result of these procedures, consumers will be able to make more informed decisions regarding home purchases, leases and maintenance to prevent lead poisoning. They can learn how to protect themselves and their families. The rule also gives anyone involved in the real estate transaction (including sellers, renters, banks, or any of their agents) flexibility to negotiate key terms of the evaluation. Sales contracts and leasing agreements must include notification and disclosure language. Sellers, lessors and real estate agents share responsibility for ensuring compliance with this regulation.

The new rule covers most public and private housing that might be occupied by children that was built before 1978, the year lead-based household paint was outlawed. It should be most important to families with children, young couples planning a family, pregnant women and people planning major renovation of older homes. The rule affects most home buyers and renters (12 million total). It applies to most private housing, public housing, federal owned housing, and housing received federal assistance.

Essentially, the EPA program is about information. It honors the right of parents to know that they are raising their children in a safe environment. The EPA recognizes that when people are informed, they can play a major role in protecting themselves and their children against environmental pollutants. With this key information on lead in their housing and the steps to take to manage the lead, parents can be a powerful force in protecting their children from lead poisoning. The rule of 1996 gives them the tools they need. An indication of the EPA's new emphasis on spreading the word about lead poisoning is the hotline it established for the public: 1-800-424-LEAD. Armed with such information and new procedures that safeguard children against lead, parents across the country can take action. The 1996 rule is a large step forward in preventing lead poisoning.[96]

Over the past century, efforts to prevent childhood lead poisoning by controlling lead-based paint hazards in housing have been marked, not by overreaction, but by subdued, halting, reactive, and, unfortunately, largely ineffective policy responses, technical guidelines, and financing. This history is due to a number of reasons, including the fact that symptoms of lead poisoning are subtle and benefits are thus intangible.

An increasing number of lawsuits brought by poisoned children against building owners has also fueled a broadly recognized need to change the way in which the nation responds to lead-based paint in housing. Recent developments suggest that the threats to children's health and the availability of older affordable housing posed by lead-based paint are finally being recognized and managed. Those developments include:[97]

    o Passage of the Residential Lead-Based Paint Hazard Reduction Act of 1992 (Title X of the 1992 Housing and Community Development Act—PL 102-550), which defines a lead-based paint hazard based on the scientific understanding of how lead exposures from paint actually occur.

    o Completion of a HUD Lead-Based Paint Task Force report mandated by Title X that represents a consensus policy statement on correcting market failures, other recommendations for private, public sector initiatives.

    o Issuance of updated HUD "Guidelines for the Evaluation and Control of Lead-Based Paint Hazards in Housing that detail specific technical procedures to control exposures from lead-based paint.

    o Funding by the federal government for hazard control in both private and public housing; as of 1996, nearly $355 million will have been appropriated at the federal level

specifically for lead-based paint hazard control in low income private housing and millions more in public housing.
o Promulgation of a new disclosure law that will provide citizens with the opportunity to find out whether or not lead-based paint may exist in their newly purchased or rented houses.
o New state and local programs aimed at making the lead-based paint hazard identification and control industry well trained and certified to protect children and make certain that owners' money is spent wisely.
o Completion of major new research projects paving the way for simpler and more inexpensive hazard identification and control methods.

Current estimates are that there are 64 million housing units with lead-based paint above one milligram per square centimeter, about half of the nation's entire housing stock. Twenty million of those units are thought to contain lead-based paint in a hazardous condition and 3.8 million children less than six years of age reside in those houses. Approximately 500,000 of those households are economically distressed and cannot finance correction of those hazards without public subsidy. Thus, lead in paint constitutes a widely dispersed, yet highly concentrated source.[103]

Previous control efforts to reduce lead exposures were more centralized and relatively inexpensive. Reduction of lead in food was accomplished primarily from the elimination of the use of lead in the seams of cans. In gasoline, reductions were made possible by changes in gasoline refining techniques. Both these measures could be implemented by addressing only a few sources (refineries and canning plants).

The fact that lead poisoning is often an asymptomatic disease has important implications for how it is viewed by policy makers and medical care providers, who may be reluctant to allocate adequate resources to control a disease that is not readily apparent. Victims may appear "normal" and compensatory, special education, nutrition, and public education programs may help to some extent. The sources of exposure, the control of which will require significant financing in a housing stock that is already distressed, adds to the perception that the lead problem is either intractable, or paradoxically, not a significant problem.

Studies of various types of lead hazard control methods and how well they protect the health of children, the public, and workers have not been compiled until recently.[104] The studies show that carefully executed hazard control methods are effective in reducing children's blood lead levels and/or the dust lead levels in their houses.

These blood lead levels appear to decline anywhere from 6 percent to 23 percent over a period of six months to a year following hazard control. Another study shows that an 84 percent to 96 percent decline in dust lead levels can be maintained for at least 3.5 years following abatement.[105]

Most of the studies were of lead-poisoned children and not explicitly designed to quantify the primary prevention benefit of controlling exposures before blood lead levels increased. Because prevention of exposure would eliminate the bone lead storage phenomenon and irreversible neurological effects, the effectiveness of hazard control will be greater than indicated by these studies.

No study has yet been done of children born into lead-safe dwellings, making an accurate quantification of the benefits of exposure prevention difficult. Indeed, such a study poses significant ethical concerns (i.e., a control group would consist of children born into houses with the lead hazards remaining untreated). But it is evident that reliance on the medical model (i.e., treatment of houses following the appearance of a child who has already been poisoned) will fail to realize the full benefits of primary prevention.

Title X of the 1992 Housing and Community Development Act charges HUD with the task of shifting the nation's strategy from a reactive, medical model approach to a preventive, housing-based approach, while maintaining the existing blood lead screening programs to ensure that those children who are poisoned are identified and treated.

During the past several years, an unnecessarily divisive debate has raged over whether abatement is more effective than interim controls, and whether public education programs can be substituted for actual correction of lead hazards. Some view interim controls to be the only practical response, given the financial condition of much of the nation's low income housing stock and the lack of funds to conduct abatement. Others view abatement to be the only effective response, since interim controls demand a heightened commitment to housing management that is unrealistic in the dwellings posing the greatest risks, where housing is operated by owners and managers who are either unwilling or unable to carry out basic management practices.

Interim controls are those methods that can be implemented quickly and relatively inexpensively, but are likely to have shorter lifespans than abatement methods. Interim controls include specialized cleaning (dust removal), paint film stabilization, friction and impact treatments, and treatment of bare soil with grass, sod or other covering. Abatement includes any treatment to a house that can be expected to last at least 20 years, including building component replacement, enclosure, paving or soil removal, encapsulation,, and paint removal.

Studies have been conducted using treatments that today would be called interim controls. Collectively, the studies demonstrate that both strategies are effective. In fact, both strategies have been integrated successfully in the nation's only truly ongoing, long-term primary prevention effort: the public housing program. Here, immediate hazards are identified and controlled, and housing units are made free of hazards and thus eligible for insurance relatively quickly and inexpensively, typically at a cost of about $200 per dwelling unit. Much of this work is done by specially trained maintenance workers who perform what the HUD Task Force termed "Essential Maintenance Practices." At the same time, long-term but deliberate progress is being made to render all such dwellings permanently lead-safe, usually through building component replacement or permanent enclosure/encapsulation in conjunction with renovation work.

Costs of lead abatement are difficult to separate from the costs of renovation. For example, replacement of old energy-inefficient windows coated with lead-based paint with new windows could be considered to be an abatement technique or renovation work. Anecdotal evidence suggests the costs of abatement are anywhere from $1,500 to $20,000 per dwelling unit, depending on the size of the house and the number of painted surfaces. In short, interim controls and abatement are complementary, not contradictory, activities.

Not surprisingly, the degree of effectiveness varies with the baseline blood lead level. It appears that the extent of the blood lead decline is most pronounced when the child's baseline blood lead level is already elevated. At lower blood lead levels, the decline is more modest, as expected. When the work is conducted without proper controls, dust lead and blood lead levels often increase, sometimes dramatically.

The harmful effects of haphazard abatement and careless home renovation or remodeling projects that disturb leadbased paint can be avoided if certain quality control, training, licensing and certification systems are developed and enforced, and if clearance examinations are conducted in dwellings following the work. Such a nation-wide system is being promulgated by EPA under statutory authority from Congress.[106] Clearance examinations include visual assessment and dust (and perhaps soil) sampling after lead hazard control work has been completed. Without such control, exposures can be heightened instead of controlled. Evidence that poorly controlled housing rehabilitation and renovation can cause lead poisoning has been reviewed elsewhere.

Lead poisoning prevention is widely regarded as one of the stellar public health initiatives of the second half of this century. While blood lead levels have declined dramatically, previous successes in controlling exposures demonstrate that such actions provide direct health benefits, not that efforts to control childhood lead poisoning are no longer needed. There are some who assert that we should stop here, that we have already done enough by controlling lead in food and gasoline. They argue that we can "coast" the rest of the way and that childhood lead poisoning will eventually disappear by itself without further actions. Paradoxically, they also argue that while needed, addressing paint hazards will cost too much.

However, Schwartz recently estimated that the nation would save $1.7 billion per year for every 0.1 microgram per deciliter drop in mean population blood lead level.[107] HUD recently completed a regulatory impact analysis for its proposed regulations covering all federally assisted housing.[108] The analysis showed that net benefits in the first year of the rule would be over $1 billion for federally assisted housing alone. The Centers for Disease Control and Prevention estimated that the nation would save $62 billion by addressing lead-based paint hazards over the next 20 years, discounted to the present.

Some suggest that the nation is about to embark on a program of complete removal of all lead-based paint from 64 million dwellings. HUD estimates that the cost of removing all lead-based paint from the nation's housing stock would cost over $500 billion and that the average cost of lead paint removal is $7,700 per housing unit.[109] Average costs for full abatement provide an incomplete picture, however, because many houses can be fully abated for a much smaller amount. In HUD's Federal Housing Administration (FHA) demonstration project, half of all the houses treated cost less than $2,500 per unit to completely abate because the number and size of surfaces coated with lead-based paint was quite small. In its lead-based paint grant program for private housing, HUD and local and state governments are experimenting with a number of techniques to reduce costs, including the use of community-based groups to do the work and better targeted innovative hazard control technologies. Many houses have only a small amount of lead-based paint present. For example, houses built between 1960 and 1979 have an average of 782 square feet of lead-based paint on interior and exterior surfaces, less than 5 percent of all paint. On the other hand, houses built before

1940 have an average of 2,355 square feet of lead-based paint on interior and exterior surfaces.[109]

Despite the fact that some houses may be candidates for fuller treatment, a $500 billion price tag indicates that complete removal of all lead-based paint in all housing is impractical, certainly in the near term. Complete removal or abatement may be feasible in houses where only a small amount of lead-based paint is present, where the incremental costs of abatement in the context of renovation are relatively inconsequential, or where funds are available. For all other houses, interim control methods aimed at controlling exposures, i.e., lead-based paint that is in a hazardous condition (along with control of contaminated house dust and bare soil) is more appropriate and feasible. In fact, there is not a single housing program anywhere that has attempted to remove all lead-based paint from all dwellings. Even the most extensive abatement programs include options such as enclosing or encapsulating lead-based paint. In addition, most housing programs, including public housing programs, have no breakneck deadline to meet for abatement. Instead, abatement activities are almost always carried out during other construction and housing renovation activities, and will be done over a decade or two in a controlled, rational fashion. This kind of deliberate effort hardly seems to be a campaign based on panic.

Interim controls and some forms of abatement are based on the idea that property owners will manage their properties in such a way that any lead-based paint that is present remains in a non-hazardous controlled condition. Are property owners in fact up to this task? Is it reasonable to require owners to take the necessary steps to keep all lead-based paint intact, provide cleanable surfaces to prevent accumulation of leaded dust, keep bare soil covered, conduct routine maintenance work in a "lead-safe" way, and conduct periodic evaluations and tests to ensure the property remains lead-safe? Some argue that owners of properties most likely to produce lead-poisoned children (dilapidated inner-city low income housing) will not, in fact, be capable of performing these management functions and that therefore full removal of lead-based paint is the only foolproof answer.[97]

What is likely to be the effect of owners' failure to adopt interim controls in a rigorous, serious fashion? Ironically, such a failure is likely to stimulate calls for wholesale removal of all lead-based paint, which is clearly impractical. After all, if it appears that owners cannot manage lead-based paint hazards, what other choice is there? The result of such a scenario is clear: calls for wholesale removal, in fact, would lead to no action at all.

Such an environment is conducive to lawsuits brought by poisoned children against landlords who failed to take action. According to a recent New York Times article, that city alone is faced with over one thousand pending lawsuits, and the tort system is equipped to address the problem. In fact, most poisoned children do not sue and of those that do, most are unsuccessful in proving damage due to the subtle effects of lead poisoning described above. Those few cases that do win substantial awards for poisoned children sometimes have the unintended consequence of driving affordable insurance and responsible owners out of the market.

Policy makers have learned useful lessons from the asbestos experience. In asbestos, EPA regulated removal operations only; no EPA regulatory apparatus was constructed to regulate or monitor asbestos as it remained in place. The lesson on the asbestos experience was not that asbestos is not a toxic substance—it is, and exposures need to be controlled so that asbestos-related disease does not occur. The lesson was simply that it is important to control exposures, which is not necessarily the same as removal of the substance. In short, some toxic substances may best be managed in place until removal becomes the only option or unless the material will be disturbed in a way that can cause hazardous exposures, such as during renovation. If management is not possible, effective, or preferred, then safe, controlled, and professional removal or abatement is the only alternative. Fundamentally, this choice is no different than any other capital improvement made in housing.

Who can and should pay for controlling childhood lead poisoning? Currently, the costs of childhood lead poisoning are borne by our nation's health care and educational systems. In spite of advances in medical treatment, complete reversal of the effects of seriously elevated blood lead levels is unlikely. Medical care costs anywhere from $1,300 to $5,000 per child[110] and many children must be treated more than once. Poisoned children will also require special attention in the educational system to attempt to overcome the reductions in intelligence caused by lead poisoning. These costs are estimated to be about $3,300 per child, a great deal more in severe cases.[110] This figure does not include the wider economic and social costs of lowered productivity, special education, substantial reductions in lifetime earnings and anti-social behavior. Spending money like this makes even less sense when some houses are known to repeatedly poison many children over many years.

A more practical, less costly (and far more humane) way to deal with this problem is to control the source of exposure. Principally, this source is lead-based paint and the contaminated dust and soil it generates. In some locations, contaminated dust and soil are also still with us from the historic use of leaded gasoline and emissions from lead industries. All these sources are found in the home, so it makes sense to focus on housing-based strategies.[97]

5. Informed Consumers and the Housing Market

Lead-based paint hazards in housing are generally not recognized or valued by the public. Housing appraisers do not consider the presence or absence of lead-based paint hazards in the value of a dwelling. Free markets require informed customers in order to operate properly. Until the housing market values a lead-safe dwelling, financing lead-based paint hazard control will not be considered to be a reasonable investment by owners. Fundamentally, the presence of lead-based paint is no different than any other housing defect, such as a leaky roof. But because the hazards are not recognized, victims are not obvious, and consumers are not informed, the market fails.

Owners do not regard lead hazard control to be a good investment because the value of the property does not increase.

HUD and EPA have issued a law that will correct this market failure.[111] The lead-based paint disclosure law permits prospective owners to obtain a lead-based paint inspection or risk assessment if they want one. It also requires owners to disclose any knowledge of lead-based paint in the dwelling before lease or sale. Finally, it will inform consumers about lead-based paint hazards through a warning statement and through dissemination of an educational brochure at approximately 9 million sales and leasing transactions annually.

Is it fair, or even practical, to expect that owners alone are responsible for paying to correct the problem? After all, most owners did not even think they were doing anything wrong in applying lead-based paint to their dwellings and in some cases they were obeying laws that required the use of lead-based paint. Furthermore, most owners today never actually applied the old paint in the buildings they now own. Many landlords who operate low income properties have very low or negative cash flows that make private financing of lead hazard controls difficult. Lead poisoning prevention efforts should not cause abandonment of housing, already a major problem. Public sector funding is needed for those economically distressed dwellings not amenable to information and market-based strategies.

Informed consumers will solve a significant part of the problem as lead hazard control becomes simply another wise investment in a house, although many low income properties do not generate enough income to make preventative measures financially sensible, even low cost interim controls. Even nonprofit housing organizations are finding it difficult to continue to provide decent housing at prices the poor can afford. In addition to its lead-based paint grant program for private housing, HUD is proposing to streamline its lead-based paint requirements across all its housing programs and to require specific action that depends on the extent of federal financial involvement.[112]

Even though subsidies and private markets will play important roles, the hard fact remains that owners control housing and are ultimately responsible for correcting any housing defects, including lead-based paint hazards. Increasingly, local laws provide a legal duty for owners to respond to hazards and poisoned children. How owners (and the insurance and mortgage companies that back them) respond to this issue will, in large measure, determine how the nation proceeds.

If owners fail to adopt a more active management style, then it is almost certain that more litigation, more high-priced awards, and more extreme calls for complete removal will arise. The net result will be that only a small fraction of poisoned children will be compensated and the largest source of lead exposure to children will be left uncontrolled. On the other hand, if owners act to implement rather simple measures that control hazards, and perhaps more importantly, if policy makers require owners to actively monitor the condition of lead-based paint to make sure hazards do not appear, then litigation will be much less successful, fewer cases will be brought, and fewer children will be poisoned.

Historically, the biggest judgments have been brought against owners who failed to do anything. Proving negligence is relatively simple in such cases. On the other hand, owners who have implemented reasonable management practices and who have data proving that the dwelling they own is lead-safe will have much less liability exposure. In this situation, the assumption will be that the child must have been poisoned from a lead source other than the paint in the dwelling.

Previous efforts to control lead in gasoline and food have been successful in reducing population blood lead levels. However, further decreases are unlikely unless additional action is implemented to control exposures to lead-based paint and the contaminated soil and dust it still generates in housing. Prevalence studies show that more pre-1978 houses contain some lead-based paint, although most paint does not, in fact, contain lead, and most of the paint that does contain lead is in a non-hazardous condition. Cost-benefit analyses show that the nation will save billions of dollars if lead-based paint hazards are controlled. HUD has recently issued comprehensive technical guidelines that govern how both interim control and abatement techniques can be implemented safely and how lead-based paint hazards can be properly identified.[112] Calls for wholesale removal of all lead-based paint are not feasible due to the high cost involved and may, in fact, result in increased exposure to children. A more flexible strategy is now being implemented that requires owners to properly manage lead-based paint using special maintenance practices and monitoring until permanent removal or enclosure/encapsulation can be performed.[112]

The principal method of implementation involves the use of existing private market forces, which require an informed consumer for most houses and targeted subsidies in cases where private financing is not possible. Through a series of regulatory and non-regulatory actions, HUD is implementing a new plan to control lead-based paint hazards in housing through a reasonable, practical strategy based on science, research, consensus and private/public partnerships. Owners have the principal responsibility to manage lead-based paint hazards in the housing they own; their response will, in large measure, determine the success or failure in containing liability exposure and the childhood lead poisoning epidemic.

In the next five to ten years, an important opportunity exists to break the back of a preventable disease that is too expensive to ignore; whose causes are well understood, whose outcome is insidious and irreversible, and whose solutions are clear.

Methods for Lead and Other Heavy Metal Testing

One method for lead/heavy metal testing is as follows:

Form an aqueous solution with pH in the range of 1-5; bring this solution into contact with the article to be tested, adding a sulfide-containing compound to the solution (this compound should be of a type which produces a colorless sulfide solution). Visually examine the resulting solution for a lead sulfide precipitate to indicate the presence or absence of lead.

The aqueous solution should have a pH of 2, and be formed from an organic acid selected from the group consisting of acetic, citric, tartaric, maleic, adipic, succinic, propionic and/or lactic. The sulfide compound should be selected from the group consisting of sulfides and hydrosulfides of lithium, sodium, potassium, ammonium, calcium, strontium, and/or barium.

Testing for Lead Using The Lead Buster™

To make the test solution, follow these steps:
1. Buy sodium sulfide in one pound quantities.
2. Mix one pound of sodium sulfide with three gallons of water.
3. Decant mixture, as there will be a slight precipitate.
4. Buy small bottles and have them appropriately labeled.
5. Fill bottles and screw on caps tightly.
(Three gallons will fill about 175 two-ounce bottles).

Your lead detection system should contain the following items:
Test tube
Test bowl
Eye dropper
File
Test solution bottle (check date before testing)

You will also need ***distilled white vinegar*** to perform the testing. It can be purchased in any grocery store.

WARNING: The test solution contains sodium sulfide, which is an irritant. Avoid contact with skin or eyes. Flush with water and wash with soap if it comes in contact with your skin; flush with water and call a physician if the solution comes into contact with the eyes.

This test gives off hydrogen sulfide gas (a rotten egg odor), so be sure to do the testing in a well-ventilated area. Some items under test will allow the vinegar to soak through. Please protect your work surfaces by putting the test item into a large plastic container, or cover your work surface so that vinegar won't damage it.

Please Note: Some states have very strict laws pertaining to lead removal if lead paint is detected in homes with children.

It is recommended that you contact your state agency before you test.

Detecting Lead In Pottery, Ceramic and Glassware:

1. Wash the item or vessel to be tested, rinse out with clean water and dry with a clean paper towel.

2. Pour enough ***distilled white vinegar*** into the item to half fill it, or to cover any colored design or decal inside the item (*Note:* drinking glasses with decals or designs on the outside should be placed upside down or on their side in a plastic bowl with enough vinegar to cover the design or decal). Cover the test item or bowl with plastic wrap to prevent evaporation and LET STAND UNDISTURBED FOR 24 HOURS.

3. After 24 hours, remove the plastic wrap. Remove the red cap from the Test Tube/ Color Chart Unit. Using the eye dropper, transfer enough test vinegar into the test tube so that it is 1/3 full. Wash and dry the dropper after use.

4. Remove the yellow cap from the test solution bottle and squeeze five (5) drops of the solution into the test tube. Replace the yellow cap on the test solution bottle; replace red cap on test tube. WHILE HOLDING THE RED CAP DOWN WITH YOUR FINGER,

SHAKE THE TEST TUBE FOR 5 SECONDS.

5. Compare liquid in test tube with the attached color chart. A white cloudy appearance is common and may indicate the presence of zinc. A yellow color indicates lead is present; a black color or precipitate indicates a high level of lead.

IMPORTANT NOTE: The color change in the solution means that lead sulfide has been formed in the test tube. Lead sulfide is poisonous! All lead sulfide should be disposed of properly.

Detecting Lead In Painted Surfaces (house paint, cribs, toys, etc.):

1. Using the file, carefully file the painted surface to be tested. A razor blade or even a sharp knife can also be used to cut paint chips. Allow the filings (or chips) to collect in your test bowl. File down to collect particles *from each layer of paint.*

2. Pour distilled white vinegar into the test bowl to cover the paint filings. Let stand for five minutes.

3. Remove the yellow cap from your test solution bottle, and squeeze five (5) drops into the test bowl and observe any color change. A black color indicates a high level of lead, 25 parts per million or more. For other values, please refer to the color chart.

An Alternative Method for Checking Painted Surfaces:

1. Instead of collecting paint filings or chips, cut a wedge out of the painted surface itself, making sure to expose every layer of paint.

2. Apply the test solution directly into the cut surface, making sure it comes into contact with each layer of paint. Two drops should be sufficient.

3. Examine the cut with a magnifying glass to determine if there has been a color change.

4. Check color change against color analysis scale, below:

FDA Color Analysis (Parts Per Million [P.P.M.] Of Lead):

Slightly Yellow*ish*	=	2 parts per million
Slight Yellow	=	4 parts per million
Definite Yellow	=	6 parts per million
Dark Yellow	=	8 parts per million
Brownish Yellow	=	12 parts per million
Heavy Black	=	25 parts per million or more

Metals such as antimony, bismuth, iron, and copper can cause color changes to occur when using a lead detection system. However, properly manufactured items should not release heavy metals in the amounts that give a color change, so it would be prudent to stop using the offending item. Items that release lead or other heavy metals should be set aside and *SHOULD NOT BE USED FOR COOKING, FOOD STORAGE OR DRINKING!*

Lead in tap water: You may have a lead problem with your tap water. Because of the low levels of lead which can be dissolved in water, some lead detection systems will not give accurate readings, making other methods necessary for this kind of lead testing.

Technical Terms, Chapter IV

encapsulation: to enclose in a membrane
heme synthesis: blood synthesizing
hyperphosphaturis: excess phosphate in the urine
mitosis: cell division
paradoxically: contrary to accepted opinion
pyknosis: cellular degeneration
uremia: excessive amounts of urea and other nitrogenous waste products in the blood

References, Chapter IV

Amitai, Y., Brown, M. J., Graef, J. W., and Cosgrove, E. (1991). "Residential Deleading: Effects on the Blood Lead Levels of Lead-Poisoned Children." Pediatrics, 88(5): 893-897.

Aschengrau, A., Beiser, A., Bellinger, A., et al. (1994). "The Impact of Soil Lead Abatement on Urban Children's Blood Lead Levels; Phase II Results from the Boston Lead-in-Soil Demonstration Project." Environmental Research, 67: 125-148.

Environmental Protection Agency. (September 2, 1994). EPA 1994: Proposed Rule, Requirements for Lead-Based Paint Activities. Federal Register, Washngton, D.C., p. 4587-45921.

Environmental Protection Agency. (May 1995). EPA 1995a: "A Field Test of Lead-Based Paint Testing Technologies." EPA 747-R-0026, Washington, D.C.

Environmental Protection Agency. EPA 1995b: "Review of Studies Addressing Lead Abatement Effectiveness." Battelle Institute, EPA 747-R-95-006.

Environmental Protection Agency. EPA 1995c: "Report on the National Survey of Lead-Based Paint in Housing." EPA 747-R-95-003.

Fett, M. J., Mira, M., Smith, J., Alperstein, G., Couser, J., Brokenshire, T., Gulson, B., Cannata, S. (October 5, 1992). "Community Prevalence Survey of Children's Blood Lead Levels and Environmental Lead Contamination in Inner Sydney." Medical Journal of Australia, 157: 441-445.

Fischbein, A., Anderson, K. K., Shigeru, S., Lilis, R., Kon, S., Sarkoi, L., and Kappas., A. (1981). "Lead Poisoning from Do-It-Yourself Heat Guns for Removing Lead-Based Paint: Report of Two Cases." Environmental Research, 24: 425-431.

Housing Environmental Services. HES 1995: Personal Communication with Miles Mahoney.

Jacobs, D. E. (1994). "Lead-Based Paint as a Major Source of Childhood Lead Poisoning: A Review of the Evidence." *Lead in*

Paint, Soil and Dust: Health Risks, Exposure Studies, Control Studies, Control Measures, Measurement Methods and Quality Assurance, edited by M. E. Beard and S. D. A. Iske. American Society for Testing and Materials STP 1126, Philadelphia, PA, p. 175-187.

Morris, R. (1995). "Why Title X Should Be Repealed." Lead Tech 1995 Conference Proceedings 43-50.

Rabinowitz, M., Leviton, A., Bellinger, D. (1985). "Home Refinishing: Lead Paint and Infant Blood Lead Levels." American Journal of Public Health, 75: 403-404.

Schwartz, J. (1994). "Societal Benefits of Reducing Lead Exposure." Environmental Research, 66: 105-124.

Shannon, M. W. and J. W. Graef. (1992). "Lead Intoxication in Infants." Pediatrics, 89(1): 87-90.

Staes, C. and Rinehart, R. (1995). "Does Residential Lead-Based Paint Hazard Control Work? A Review of the Scientific Evidence." National Center for Lead-Safe Housing, Columbia, Maryland, 79 pages.

Swindell, S., Chancy, E., Brown, M. J., Delaney, J. (1994). *Home Abatement and Blood Lead Level Changes in Children with Class III Lead Poisoning.*

U.S. Department of Housing and Urban Development. HUD 1990: "Comprehensive and Workable Plan for the Abatement of Lead-Based Paint in Privately Owned Housing." Report to Congress, Washington, D.C.

U.S. Department of Housing and Urban Development, Office of Policy Development and Research. HUD 1991: "The HUD (FHA) Lead-Based Paint Abatement Demonstration Project." Prepared by Dewberry & Davis, HC-5831, Washington, D.C.

U.S. Department of Housing and Urban Development, Office of Lead-Based Paint Abatement and Poisoning Prevention. (June 1995). HUD 1995a: "Lead-Based Paint Hazard Reduction and Financing Task Force—Putting the Pieces Together: Controlling

Lead Hazards in the Nation's Housing." HUD-1547-LBP, Washington, D.C.

U.S. Department of Housing and Urban Development, Office of Lead-Based Paint Abatement and Poisoning Prevention. HUD 1995b: "Guidelines for the Evaluation and Control of Lead-Based Paint Hazards in Housing." HUD-1539-LBP, Washington, D.C.

U.S. Department of Housing and Urban Development, Office of Lead-Based Paint Abatement and Poisoning Prevention. HUD 1995c: "Lead-Based Paint Hazard Control Grant Program for Low-Income Private Housing." Washington, D.C.

U.S. Department of Housing and Urban Development, Office of Lead-Based Paint Abatement and Poisoning Prevention. HUD 1995d: "Regulatory Impact Analysis of the Proposed Rule on Lead-Based Paint: Requirements for Notification, Evaluation and Reduction of Lead-Based Paint Hazards in Federally-Owned Residential Property and Housing Receiving Federal Assistance." ICF Kaiser, Washington, D.C.

U.S. Department of Housing and Urban Development, Office of Lead-Based Paint Abatement and Poisoning Prevention. (March 1996). HUD 1996: "Interim Report on the Evaluation of Controlling Hazards in Low-Income Housing." The National Center for Lead-Safe Housing and the University of Cincinnati.

U.S. Department of Housing and Urban Development and the Environmental Protection. (March 6, 1996). HUD/EPA 1996: "Joint Rule, Requirements for the Disclosure of Known Lead-Based Paint and/or Lead-Based Paint Hazards in Housing." Federal Register, p. 9063-9088.

ENDNOTES

[1] Nriaga, J.O. "Lead and Lead Poisoning in Antiquity." *Environmental Science and Technology Series.* (NY: John Wiley & Sons, 1983).

[2] Hughes, J.T., Horan, J.J. & Powles, C.P. (1976, October 13). "Lead Poisoning Caused by Glazed Pottery." Case Report: *New Zealand Medical Journal*, 13 October 1976: 266-68.

[3] Klein, M., Namer, R., Harpur, E. & Corbin, R. "Earthenware Containers as a Source of Fatal Lead Poisoning: Case Study and Public Health Considerations." *New England Journal of Medicine*, 1970. 283: 669-72.

[4] FDA Consumer Memo. "Glazes and Decals on Dinnerware." U.S. Department of Health, Education & Welfare, Public Health Service, Food & Drug Administration. October 1979 (Revised).

[5] "Facts About Lead Glazes for Art Potters and Hobbyists." (NY: Lead Industries Association, Inc., 1972).

[6] Arena, J.M. *Poisoning: Toxicology, Symptoms, Treatments. Fourth Edition.*
(Springfield, IL: Charles C. Thomas, Publisher, 1978): 255-56.

[7] Mahaffey, K.R., Annest, J.L., Roberts, J. & Murphy, H.S. "National Estimates of Blood Lead Levels, United States, 1976-1980:

Association with Selected Demographic and Socioeconomic Factors." *New England Journal of Medicine*, 1982. 307:573-79.

[8] Natelson, E.A. & Fred, H.L. "Lead Poisoning from Cocktail Glasses: Observations on Two Patients. *Journal of the American Medial Association*, 1976. 236:252-57.

[9] Whitehead, T.P. & Prior, A.P. "Lead Poisoning from Home-made Wine." *The Lancet*, 17 December 1976. 2:133-34.

[10] Williams, M.K. "Lead Poisoning: An Unusual Complication of Cystitis." *The Lancet*, 23 September 1972. 2:662-63.

[11] Tavolato, B., Lieandro, A.C. & Argentiero, V. "Lead Polyneuropathy of Nonindustrial Origin." *Eur. Neurol.*, 1980. 19:273-76.

[12] Barry, P. "Lead-Glazed Earthenware." *The Lancet*, 23 September 1972. 23:662-63.

[13] Harris, R.W. & Elsea, W.R. "Ceramic Glaze as a Source of Lead Poisoning." *Journal of the American Medial Assoc.*, 1967. 202:544-46.

[14] Bird, T.D., Wallace, D.M. & Labbe, R.F. "The Porphyria, Plumbism, Pottery Puzzle." *Journal of the American Medical Assoc.*, 1982. 247:813-14.

[15] Miller, C. "The Pottery and Plumbism Puzzle." *The Medical Journal of Australia*, October 1982. 30:442-43.

[16] Lee, W.R. "What Happens in Lead Poisoning?" *Journal of the Royal College of Physicians of London*, 1981. 15:48-54.

[17] Fischbein, A., Anderson, K.E., Sassa, S., et al. "Lead Poisoning from Do It Yourself" Heat Guns for Removing Lead-Based Paint: Report of Two Cases." *Environmental Research*, 1981. 24:425-31.

[18] Wessel, M.A. & Dominski, A. "Our Children's Daily Lead." *American Scientist*, 1977. 65:942-46.

[19] Dreisbach, R.H. *Handbook of Poisoning: Prevention, Diagnosis and Treatment, Eleventh Edition.* (Los Altos, CA: Lange Medical Publications, 1983): 252-60.

[20] Cullen, M.R., Robins, J.M. & Eskenazi, B. "Adult Inorganic Lead Intoxication: Presentation of 31 New Cases and a Review of Recent Advances in the Literature." *Medicine*, 1983. 62:221-47.

[21] Kehoe, R.A. "The Metabolism of Lead in Health and Disease." *Journal of the Royal Institute of Public Health*, 1961. 24:1-81, 101-120, 129-43, 177-203.

[22] Lin-Fu, J.S. "Undue Absorption of Lead Among Children: A New Look at an Old Problem." *The New England Journal of Medicine*, 1972. 286:707-710.

[23] U.S. Food and Drug Administration. "Advanced Notice of Proposed Rulemaking: Request for Data." *The Federal Register*, 1979. 44(171):51233-242.

[24] Hawkins, H. "Lead—A Weighty Problem." *Fd. Cosmet. Toxicol.*, 1979. 17:171-72.

[25] Biddle, G.N. "Toxicology of Lead: Primer for Analytical Chemists." *Journal of the Assoc. Of Official Analytical Chemists*, 1982. 65:947-52.

[26] Settle, D.M. & Patterson, C.C. "Lead in Albacore: Guide to Lead Pollution in Americans." *Science*, 1980. 207:1167-76.

[27] Goldberg, A. & Beattie, A.D. "Role of Lead in Medicine." *Recenti Progressi In Medicina*, 1974. 56:497-519.

[28] Gilfillan, S.C. "Lead Poisoning and the Fall of Rome. *Journal of Occupational Medicine*, 1965. 7:53-60.

[29] Neuhauser, W. "Caesar's Teeth and Lead: Causes of Toothlessness and Infertility in Ancient Rome." *Quintessenz*, 1975. 26:107-09.

[30] Lacambre, O. "Lead Poisoning and the Roman Empire." *Ann. Hyg. Lang.* France, 1975. 11:29-31.

[31] Nriagu, J.O. "Saturnine Gout Among Roman Aristocrats: Did Lead Poisoning Contribute to the Fall of the Roman Empire?" *The New England Journal of Medicine*, 1983. 308:660-63.

[32] McCord, C.P. "Lead and Lead Poisoning in Early America: Benjamin Franklin and Lead Poisoning." *Industrial Medicine and Surgery*, 1953. 22(9):393-99.

[33] United Press International "300 Get Lead Poisoned." *The Seattle Times*, 12 December 1983, sec. A:4.

[34] McCabe, E.B. "Age and Sensitivity to Lead Toxicity: A Review." *Environmental Health Perspectives*, 1979. 29:29-33.

[35] Watson, W.S. "Oral Absorption of Lead and Iron." *The Lancet*, 2 August 1980. 2:236-37.

[36] Corwin, E. "On Getting the Lead Out of Food." *FDA Consumer,* March 1982. 19:11-13.

[37] Jelinek, D.F. "Levels of Lead in the United States Food Supply." *Journal of the Assoc. of Official Analytical Chemists,* 1982. 65:942-46.

[38] Bander, L.K., Morgan, K.J. & Zabik, M.E. "Dietary Lead Intake of Preschool Children." *American Journal of Public Health,* 1983. 73:789-94.

[39] Krinitz, B. & Hering, R.K. "Toxic Metals in Earthenware." *U.S. Government Printing Office,* Reprint from FDA Papers, April 1971. 435-654/53.

[40] U.S. Food & Drug Administration. *Compliance Policy Guide 7117.07: Ch. 17—Food Related. Subject: Pottery (Ceramics) Imported & Domestic: Lead Contamination.*
Issuing Office: EDRO, Division of Field Regulatory Guidance. Authority: Associate Commissioner for Regulatory Affairs, 1 October 1980.

[41] Horowitz, William (Ed). *Cadmium and Lead in Earthenware (8), Official Final Action, AOAC-ASTM Method 25.031-25.034* in Ch. 25, "Metals and Other Elements as Residues in Foods." *Official Methods of Analysis of the Assoc. Of Official Analytical Chemists, 13th. Edition,* (Washington D.C.): 1980.

[42] Seth, T.D., Sircar, S. & Hasan, M.Z. "Studies on Lead Extraction from Glazed Pottery Under Different Conditions." *Bulletin of Environmental Contamination & Toxicology,* 1973. 10:51-56.

[43] Henderson, R.W., Andrews, D. & Lightsey, G.R. "Leaching of Lead from Ceramics." *Bulletin of Environmental Contamination & Toxicology*, 1979. 21:102-04.

[44] Acra, A., Dajani, R., Raffoul, Z. & Karahagopian, Y. "Lead-Glazed Pottery: A Potential Health Hazard in the Middle East." *The Lancet*, 21 February 1981. 1:433-34.

[45] Koplan, J.P., Wells, A.V., Diggory, H.J.P. & Baker, E.L. "Lead Absorption in a Community of Potters in Barbados." *International Journal of Epidemiology*, 1977.
6(3):225-29.

[46] Associated Press: "N.Y. Health Department Warns of Dangerous Dinnerware." *The Seattle Times*, 6 May, 1981, sec. E:7.

[47] Molina-Ballesteros, G., Zuniga-Charles, M.A., Ortega, A.C., et al. "Lead Concentrations in the Blood of Children from Pottery-Making Families Exposed to Lead Salts in a Mexican Village." *Bulletin of the Pan American Health Organization*, 1983. 17(1):35-41.

[48] Goldfield, M. & Altman, R. "Lead Absorption from Imported Pottery—New Jersey." *Epidemiologic Notes and Reports, Morbidity and Mortality Weekly Report*, 10 August 1974. 23:284.

[49] Boling, Rick. "Heavy Metal: Weighing the Dangers of Lead in Your Dinnerware."
Harrowsmith, November 1986 (Camden East, Ontario, Canada).

[50] Veisskopf, Michael "Lead Astry: The Poisoning of America." *Discover*, December 1987.

[51] Sugarman, Carole. "Plates That Bombed—Fiesta: Red Hot Tableware." *Harrowsmith*, November 1986 (Camden East, Ontario, Canada).

[52] "Prop. 65 Warnings for Mercury in Dental Fillings." Environmental Law Foundation News Release, 14 December 1993.

[53] "Dentists to Post Warnings on Mercury." *San Francisco Examiner*, 15 December 1993, sec. A:1.

[54] "California Dentists Now Required to Inform Patients of Risks of Toxic Substances in Dental Fillings." *Townsend Letter for Doctors*, April 1994. #129 (Pt. Townsend, WA).

[55] Vimy, M.J. & Lorscheider, F.L. "Serial Measurements of Intra-Oral Air Mercury: Estimations of Dose from Dental Amalgams." *J. Dent. Res*, August 1985. 64:1075.

[56] Gay, Cox, & Reinhard."Chewing Releases Mercury From Fillings." *The Lancet*, 5 May 1979. 985.

[57] Queen, H.L. *Chronic Mercury Toxicity: New Hope Against an Endemic Disease*. (Queen & Company Health Comm., Inc. Colorado Springs, CO, 1988):7.

[58] Virkkunen, M. "Behavioural Changes Relating to Changes in Total Serum Cholesterol," *JAMA* 1 Feb 1985.

[59] Eggleston, D.W. & Nylkander, M. "Correlation of Dental Amalgam with Mercury in Brain Tissue." *Journal of Prosthetic Dentistry*, December 1987. 58:6.

[60] Eyl, Thomas R. "Methyl Mercury Poisoning in Fish and Human Beings." *Modern Medicine*, 16 November 1970.

[61] Amin-Zaki, L., et al. "Intra-Uterine Menthyl Mercury Poisoning in Iraq." *Pediatrics*, November 1974. 54(5):587-595.

[62] Queen H. L. *Mercury Free News*, May 1992. 5(2):10.

[63] Corbin, S.B. & Kohn, W.G. "The Benefits and Risks of Dental Amalgam: Current Findings Reviewed." *Journal of the American Dental Association*, April 1994. 125:383.

[64] Adam, R.D. & Victor, M. *Principles of Neurology, Fifth Edition* (1993).

[65] Ziff, S. "Silver Dental Fillings: The Toxic Time Bomb." *Aurora Press* 1984. p. 8-9.

[66] Huggins Diagnostic Center. *Position Papers*: "Amalgam Issue, Root Canals, Cavitations." p. 15.

[67] Brune, C., et al. "Gastrointestinal and In Vitro Release of Copper, Cadmium, Indium and Zinc from Conventional and Copper-Rich Amalgam." *Scand. J. Dent. Res.*, 1983. 91:66-71.

[68] Huggins, H.A. *It's All In Your Head: Diseases Caused by Silver-Mercury Filling.* (Life Sciences Press, 1989): 28.

[69] Placidi, G.F., et al. "Distribution of Inhaled Mercury (203Hg) in Various Organs." *Int. J. Tiss. React.*, 1983. 5:193-200.

[70] Trakenberg, J.M. *Chronic Effects of Mercury on Organisms.* Ch.VI: "The Micro mercurialism Phenomenon in Mercury Handlers" DHEW Publ. 1974. No (NIH) 74:473.

[71] Kasper, E.K., et al. "The Causes of Dilated Cardiomyopathy: A Clinicopathologic Review of 673 Consecutive Patients." *JACC*, 1 March 1994. 23(3):386-90.

[72] Queen, J.L. *Heart Talk*, Feb. 1989 8(1):5.

[73] Shiraki, H. & Nagashima, K. in *Neurotox*. (Roizin L. et al, Eds, 1977: pp 247-60) "Essential Neuropathology of Alkyl mercury Intoxications in Humans from the Acute to the Chronic Stage with Special Reference to Experimental Whole Body Autoradiographic Study Using Labeled Mercury Compounds."

[74] Weening, J.J. in Van dan Denergen (1980). Chapter 4: "Mercury Induced Immune Complex Glomerulopathy: An Experimental Study".

[75] Huggins, H.A. *The Applications Textbook* (1988), p 365.

[76] Huggins, H.A. (unpublished data).

[77] Louria, D.B. "Trace Metal Poisoning" in *Cecil Textbook of Medicine* (Wyngaarden & Smith, Eds.) W.B. Saunders Co., 1985.

[78] Kennedy, V.C. "Geochemical Studies in the Coeur d'Arlene Mining District, Idaho." *Geol. Survey* Circ. 168, 1952.

[79] Powers, Harold et al. "Geophysical Case History, Fredericktown Lead District, Missouri." *Min. Eng.*, March 1953. 5(3):317-20.

[80] Huff, Lyman C. "Abnormal Copper, Lead, and Zinc Content of Soil Near Metalliferous Veins." *Econ. Geol.*, August 1952. 47(5):517-542.

[81] Huttl, John B. "American Smelting & Refining Company's Van Stone Mine." *Eng. Min. Jour.*, April 1953. 154(4):72-76.

[82] Kenworthy, H., Calhoun, W.A., & Fine, M.M. "Investigation of Concentration Sections at the Central Mill of the Eagle Picher Mine & Smelting Company, Cardin, OK." Bureau of Mines *Report of Investigations*, 1949. 4511:37.

[83] "What's New In Lead Smelting?" *Mining World*, Feb. 1955. 17(2):44-46.

[84] Morgan, S.W.K. "The Production of Zinc in Blast Furnace." *Bull Inst. Min. & Met.* (England) August 1957. No. 609, Trans., 66(II):550-65.

[85] Davey, T.R.A. "Debismuthinizing of Lead." *Jour. Metals*, Mar. 1956. 3(3):341-50.

[86] "Lead in Modern Industry." Lead Industries Association, 1952.

[87] "Mineral Facts & Problems." Staff of Bureau of Mines, 1960. pp 511-19.

[88] Mentch, Robert L. & Lansche, Arnold M. "Cadmium: A Materials Survey." Bureau of Mines, 1959. Int. Circ. 7881.

[89] Langbein, G. & Brannt, W.T. *Electro-Deposition of Metals.* (NY: Henry Carey Baird & Company, Inc., 1920):335-340.

[90] US HEW OSHA "Occupational Exposure to Lead: Final Standard." *Federal Register*, 14 Nov 1978: 52953-53014.

[91] World Health Organization (WHO). *Environmental Health Criteria 3: Lead.* (Geneva: WHO, 1977).

[92] Baker, Goyer, Fowler, et al. "Occupational Renal Exposure: Nephropathy & Renal Cancer." *American Journal of Industrial Medicine*, 1980. 1:139-48.

[93] Piomelli, Seaman, Zullow et al. "Threshold for Lead Damage to Hemesynthesis in Urban Children." *Proc. National Academy of Science*, 1982. 79:3335-39.

[94] Baker, Landrigan, Barbour, et al. "Occupational Lead Poisoning in the United States: Clinical & Biochemical Findings Related to Blood Levels." *Br. J. Ind. Med.*, 1979. 36:314-22.

[95] Browner, Carol M. *Lead Perspective*. (US Environmental Protection Agency). October 1996: p 16.

[96] Goldman, Lynn R. MD. *Lead Perspective*. (US Environmental Protection Agency) October 1996: p. 17.

[97] Jacabs, David E., CIH. US HUD, Office of Lead-Based Paint Abatement & Poisoning Prevention. *Lead Perspective*. October 1996: pp 8-13.

[98] Wedeen, R.P. *Poison in the Pot: The Legacy of Lead*. (IL: Southern Illinois University Press, 1984).

[99] Chisholm, J.J. "Current Status of Lead Poisoning in Children." *Southern Medical Journal*, 1976. 69:529.

[100] Fischbein, A. "Environmental and Occupational Lead Exposure" in Rom, W.H. (Ed) *Environmental and Occupational Medicine* (Boston: Little, Brown, 1983).

[101] Piomelli, S. "Chemical Toxicity of Red Cells." *Environmental Health Perspective*, 1981. 39:65-70.

[102] Wrenn, M.E. & Mays, C.W. "Characteristics of Ionizing Radiation." in Rom, W.H. (Ed) *Environmental and Occupational Medicine* (Boston: Little, Brown, 1983).

[103] HUD 1995a US Dept of Housing & Urban Development (HUD) Office of Lead-Based Paint Abatement & Poisoning Prevention, Lead Based Paint Hazard Reduction & Financing Task Force. "Putting the Pieces Together: Controlling Lead Hazards in the Nation's Housing." (HUD-1547, LBP: Washington DC, June 1995).

[104] Staes, C & Rinehart, R. *Does Residential Lead-Based Paint Hazard Control Work?" A Review of the Scientific Evidence*, (Columbia, MD: National Center for Lead-Safe Housing, 1995).

[105] Farfel, M.R., Chisolm, J.J., & Rohde, C.A. "The Longer-Term Effectiveness of Residential Lead Paint Abatement." *Environmental Research*, 1994. 66:217-221.

[106] EPA. "Proposed Rule, Requirements for Lead Based Paint Activities." *The Federal Register*, Washington DC, 2 September 1994. 45871-45921.

[107] Schwartz, J. "Societal Benefits of Reducing Lead Exposure." *Environmental Research*, 1994. 66:105-124.

[108] HUD 1995d, Office of Lead-Based Paint Abatement and Poisoning Prevention. "Regulation of the Proposed Rule on Lead-Based Paint: Requirements for Notification, Evaluation and Reduction of Lead-Based Paint Hazards in Federal Property and Housing Receiving Federal Assistance." ICF Kaiser, Washington, DC.

[109] US Dept. Of Housing & Urban Development. "Comprehensive and Workable Plan for the Abatement of Lead-Based Paint in Housing." Report to Congress, 1990. Washington, DC.

[110] *Strategic Plan for the Elimination of Childhood Lead Poisoning.*)Atlanta, GA: Center for Disease Control & Prevention, Public Health Service, Department of Health & Human Services, 1991).

[111] "Requirements for the Disclosure of Know Lead-Based Paint and/or Lead-Based Paint Hazards in Housing." *US Dept. of HUD, and EPA Joint Rule*, 1996:9063-9088.

[112] National Center for Lead-Safe Housing & University of Cincinnati."Interim Report on the Evaluation of Controlling Hazards in Low Income Private Housing." US Dept of HUD, Lead-Based Paint Abatement & Poisoning Prevention, 1996.

[113] EPA "Quality Criteria for Atmospheric Lead." *Federal Register*, 22 August 1977. (77-24263): 42(162).

BIBLIOGRAPHY

Adam, R.D., & Victor, M. (1993). *Principals of Neurology. (5th edition)*.

Acra, A., Dejani, R., Raffoul, Z.& Karahagopian (21 Feb 1981). Lead-glazed Pottery: A Potential Health Hazard in the Middle East. *The Lancet, 1*, 433-34.

Amitai, Y. Brown, M.M., Graef, J.W., & Cosgrove, E. (November 1974). Residential Deleading: Effects on the Blood Lead Levels of Lead-Poisoned Children. *Pediatrics*, 54(5):587-595.

Arena, J.M. (1978). *Poisoning: Toxicology, Symptoms, Treatments. 4th Edition*. Springfield, IL: Charles C. Thomas, Publisher.

Aschengrau, A., Beiser, A., & Bellinger, A. (1994). The Impact of Soil Lead Abatement on Urban Children's Blood Lead levels: Phase II Results from the Boston Lead-In-Soil Demonstration Project. *Environmental Research*, 67:1235-48.

Baker, E.L., Goyer, R.A., Fowler, B.A., Knetty, U., Bernard, D.B., Alder, S., White, R.de V., Babyan, R. & Feldman, R.G. (1980). Occupational Renal Exposure, Nephropathy and Renal Cancer. *American Journal of Industrial Medicine*, 1:139-48.

Baker, E.L., Landrigan, P.J., Barbour, A.G., Cox, D.H., Folland, D.S., Ligo, R.N. & Throcknorton, J. (1979). Occupational Lead-

Poisoning in the United States: Clinical and Biochemical Findings Related to Blood Levels. *Am. Journal of Industrial Medicine*, 36:314-22.

Bander, L.K., Morgan, K.L., & Zabik, M.E. (1983). Dietary Lead Intake of Preschool Children. *American Journal of Public Health*, 73:789-94.

Barry, P. (23 September 1972). Lead-Glazed Earthenware. *The Lancet*, pp. 662-63.

Biddle, G.N. (1982). Toxicology of Lead: Primer Analytical Chemists. *Journal of the Association of Official Analytical Chemists*, 65:947-52.

Bird, T.D., Wallace, D.M. & Labbe, R.F.(1982). The Porphria, Plumbism Pottery Puzzle. *Journal of the American Medical Association*, 247:813-14.

Boling, R. (November 1986). Heavy Metal: Weighing the Dangers of Lead in Your Dinnerware. *Harrowsmith*.

Browner, C.M. (October 1996). U.S. Environmental Protection Agency: *Lead Perspectives, 16.*

Brune, C. (1983). Gastrointestinal and In Vitro Release of Copper, Cadmium, and Zinc from Conventional and Copper-Rich Amalgam. *Scandinavian Journal of Dental Research*, 91:66-71.

Bureau of Mines, U.S. Dept. Of the Interior (1960). *Mineral Facts and Problems*. Washington, DC: U.S. Printing Office.

Cadmium and Lead In Earthenware (8) (1980). Official Final Action, AOAC-ASTM Method 25.031-25.034. Chapter 25, Metals and Other Elements as Residues in Foods in *Official Methods of Analysis of the Association of Official Analytical Chemists, 13th Edition* (Wm. Horowitz, Ed.) Washington, D.C.

Choi, B.H. (1986). Methylmercury Poisoning of the Developing Nervous System: Pattern of Neuronal Migration in the Cerebral Cortex. *Neurotoxicology*, 7(2):591-600.

Corbin, S.B. & Kohn, W.G. (April 1994). The Benefits and Risks of Dental Amalgam: Current Findings Reviewed. *Journal of the American Dental Association*, 125:383.

Corwin, E. (March 1982). On Getting Lead Out of Food. *FDA Consumer*, 19:11-13.

Cullen, M.R., Robins, J.M. & Eskenazi, B. (1983) Adult Inorganic Lead Intoxication: Presentation of 31 New Cases and A Review of Recent Advances in the Literature. *Medicine*, 62:211-47.

Davey, T.R.A. (March 1956). Debismuthinizing of Lead. *Journal of Metals*, 8(3):341-350.

Driesbach, R.H. (1983). *Handbook of Poisoning: Prevention, Diagnosis and Treatment, 11th Edition*. Los Altos, CA: Lange Medical Publications.

Eggleston, D.W. & Nylander, M. (December 1987). Correlation of Dental Amalgam with Mercury in Brain Tissue. *Journal of Prosthetic Dentistry*, 58(6):704-07.

Environmental Law Foundation (14 December 1993). News Release. "Proposition 65 Warnings for Mercury in Dental Fillings."

Environmental Protection Agency (2 September 1994). EPA 1994: Proposed Rule Requirements for Lead-Based Paint Activities. *Federal Register*, 4587-45921.

Environmental Protection Agency (1995). EPA 1995a: A Field Test For Lead-Based Paint Testing Technologies. EPA 747-R-0026. Washington, D.C.

Environmental Protection Agency (1995). EPA 1995b: Review of Studies Addressing Lead Abatement Effectiveness. EPA 747-R-95-006. Battelle Institute.

Environmental Protection Agency (1995). EPA 1995c Report on The National Survey of Lead-Based Paint in Housing. EPA 747-R-95-003.

Eyl, Thomas B. (16 November 1970). Poisoning in Fish and Human Beings. *Modern Medicine*, 135-141.

Facts About Lead Glazes for Art Potters and Hobbyists. New York, NY: Lead Industries Association, Inc., 1972).

Fett, M.J., Mira, M, Smith, J., Alperstein, G., Couser, J., Brokenshire, T., Gulson, B., & Cannata, S. (5 October 1992). Community Prevalence Survey of Children's Blood Lead Levels and Environmental Lead Contamination in Inner Sydney. *Medical Journal of Australia*, 157:441-45.

Fischbein, A., Anderson, K.E. & Sassa, S. (1981). Lead Poisoning from "Do-It-Yourself" Heat Guns for Removing Lead-Based Paint: Report of Two Cases. *Environmental Research*, 24:425-31.

Food & Drug Administration (1979). Advanced Notice of Proposed Rulemaking: Request for Data. *Federal Register*, 44(171):51233-242.

Gay, Cox, & Reinhard (5 May 1979), Chewing Releases Mercury From Fillings. *Lancet*, 985.

Gilfillian, S.C. (1965). Lead Poisoning and the Fall of Rome. *Journal of Occupational Medicine*, 7:53-60.

Goldberg, A. & Beattie, A.D. (1974). The Role of Lead in Medicine. *Recenti Progressi In Medicina*, 56:497-519.

Goldfield, M. & Altman, R. (10 August 1974). Lead Absorption from Imported Pottery—New Jersey. *Epidemiologic Notes and Reports, Morbidity and Mortality Weekly Report*, 23:284.

Goldman, Lynn R., MD (October 1996). U.S. Environmental Protection Agency. *Lead Perspectives*, 16.

Harris, R.W. & Elsea, W.R. (1967). Ceramic Glaze as a Source of Lead Poisoning. *Journal of the American Medical Association*, 202:544-46.

Hawkins, R. (1979). Lead—A Weighty Problem. *Fd. Cosmet. Toxicol.* 17:171-72.

Henderson, R.W., Andrews, D., & Lightsey, G.R. (1979). Leaching of Lead from Ceramics. *Bulletin of Environmental Contamination and Toxicology*, 21:102-04.

Huff, Lyman C. (August 1972). Abnormal Copper, Lead and Zinc Content of Soil Near Metalliferous Veins. *Econ. Geol.* 47(95):517-42.

Huggins, H.A. (1993). *It's All In Your Head: Diseases Caused by Silver-Mercury Fillings.* Life Sciences Press.

Huggins, H.A. (1993). *It's All In Your Head: The Link Between Mercury Amalgams and Illness.* Paragon Press.

Huggins, H.A. (1988). *The Applications Textbook.*

Hughes, J. T., Horan, J.J. & Powles, C.P. (13 October 1976). Lead Poisoning Caused by Glazed Pottery: Case Report. *New Zealand Medical Journal.*

Huttl, John B. (April 1953). American Smelting and Refining Company's Van Stone Mine. *Engineering Mining Journal,* 154(4).

Jacobs, David E. (October 1996). CIH, HUD Office of Lead-Based Paint Abatement and Poisoning Prevention. *Lead Perspectives.*

Jacobs, David E. (1994). Lead-Based Paint as a Major Source of Childhood Lead Poisoning: A Review of Evidence. *Lead in Paint, Soil, and Dust: Health Risks, Exposure Studies, Control Studies, Control Measures, Measurement Methods and Quality Assurance.* Philadelphia, PA: American Society for Testing and Materials. STP 1126.

Jelinek, D.F. (1982). Levels of Lead in the United States Food Supply. *Journal of the Association of Official Analytical Chemists,* 65:942-46.

Kasper, E.K. (1 March 1994). The Causes of Dilated Cardiomyopathy: A Clinico-pathologic Review of 653 Consecutive Patients. *JACC,* 23(3):586-590.

Kehoe, R.A. (1961). The Metabolism of Lead in Health and Disease. *Journal of the Royal Institute of Public Health, 24.*

Kennedy, V.G. (1952). Geochemical Studies in the Coeur d''Arlene Mining District Idaho. *Geol. Survey,* Circ. 168.

Kenworthy, H. Calhoun, W.A. & Fine, M.M. (1949). Investigation of Concentration Sections at the Central Mill of the Eagle-Picher Min. & Smelting Co., Cardin, OK. *Bureau of Mines Report of Investigations.* 4511:37.

Klein, M, Namer, R., Harpur, E. & Corbin, R. (1976). Earthenware Containers as a Source of Fatal Lead Poisoning (Case Study and Public Health Considerations). *New England Journal of Medicine,* 283:669-72.

Koplan, J.P., Wells, A.V., Diggory, H.J.P., & Baker, E.L. (1977). Lead Absorption in a Community of Potters in Barbados. *International Journal of Epidemiology.* 6(3):225-29.

Krinitz, B. & Hering, R.K. (April 1971). *Toxic Metals in Earthenware.* Reprint from FDA Papers, U.S. Government Printing Office, 435-645/53.

Kuntz, W.D. (15 June 1982). Maternal and Cord Blood Background Mercury Levels: A Longitudinal Surveillance. *American Journal of Obstetrics and Gynecology,* 143:440-443.

Lacambre, O. (1975). Lead Poisoning and the Roman Empire. *Ann. Hyg. Lang.* (France). 11:29-31.

Langbein, G. & Brannt, W.R. (1920). *Electo-Deposition of Metals.* NY: Henry Carey Baird & Company, Inc.

Lead Industries Association (1952). *Lead in Modern Industry.*

Lee, W.R. (1981). What Happens in Lead Poisoning? *Journal of the Royal College of Physicians of London*, 15:48-54.

Lin-Fu, J.S. (1972). Undue Absorption of Lead Among Children: A New Look at an Old Problem. *The New England Journal of Medicine*, 286:707-710.

Louria, D.B. (1985) Trace Metal Poisoning in *Cecil Textbook of Medicine*. Wyngaarden & Smith (Eds.). W.B. Saunders Company.

MacMillan, D.L.(1977). *Mental Retardation in School and Society*. Boston: Little, Brown, & Company.

Mahaffey, K.R., Annest, J.L. Roberts, J., & Murphy, R.S. (1982). National Estimates of Blood Lead Levels, United States 1976-1980, Association with Selected Demographic and Socioeconomic Factors. *The New England Journal of Medicine*, 307:573-79.

Marlowe, M. (1984). Main and Interaction Effects of Metal Pollutants in Emotionally Disturbed Children, in *Monograph in Behavioral Disorders*, R.B. Rutherford, Ed. Reston, VA: Council for Children with Behavioral Disorders, pp. 67-79.

McCabe, E.B. (1979). Age and Sensitivity to Lead Toxicity: A Review. *Environmental Health Perspectives*, 29:29-33.

McCord, C.P. (1953). Lead and Lead Poisoning in Early America: Benjamin Franklin and Lead Poisoning. *Industrial Medicine and Surgery*, 22(9):393-99.

Meining, G.E. (1993). *Root Canal Cover-up Exposed!* Ojai, CA: Bion Publishing.

Metch, R.L. & Lansche, A.M. (1959) Cadmium: A Materials Survey. *Bureau of Mines*, Int. Circ. 7881:43.

Miller, C. (30 October 1982). The Pottery and Plumbism Puzzle. *The Medical Journal of Australia.*

Mineral Facts and Problems. (1960). Staff of Bureau of Mines, Eds.

Mining World (February 1955). What's New in Lead Smelting, 17(2):44-48.

Molina-Ballesteros, G., Zuniga-Charles, M.A. & Ortega, A.C. (1983). Lead Concentrations In the Blood of Children from Pottery-Making Families Exposed to Lead Salts in a Mexican Village. *Bulletin of the Pan American Health Organization*, 17(1):35-41.

Morgan, S.W.K. (August 1957). The Production of Zinc in a Blast Furnace. Bull
Industrial Mining and Metals (England). No. 609, Trans., 66 (Part II), 553-565.

Natelson, E.A. & Fred, H.L. (1976). Lead Poisoning from Cocktail Glasses: Observations on Two Patients. *Journal of the American Medical Association*, 236:2527.

National Institute for Occupational Safety & Health, Public Health Service Center for Disease Control, U.S. Dept. Of Health, Education & Welfare (1973). *The Industrial Environment—Its Evaluation and Control.* Washington, D.C.: Supt. Of Documents, U.S. Printing Office.

Neuhauser, W. (1975). Ceasar's Teeth and Lead: Causes of Toothlessness and Infertility in Ancient Rome. *Quintessenz*, 26:107-09.

Nriagu, J.O. (1983). *Lead and Lead Poisoning in Antiquity.* Environmental Science and Technology Series. NY: John Wiley & Sons.

Nriagu, J.O. (1983). Saturnine Gout Among Roman Aristocrats: Did Lead Poisoning Contribute to the Fall of the Roman Empire? *The New England Journal of Medicine*, 308:660-63.

Piomelli, S., Seaman, C., Zullow, D., Curran, A. & Davidow, B. (1982). Threshold for Lead Damage to Heme Synthesis in Urban Children. *Proc. Nat'l Academy of Science*, 79:3335-39.

Placid, G.F. (1983). Distribution of Inhaled Mercury (203 Hg) in Various Organs. *Internat'l Journal of Tissue Reactions*, 5:193-200.

Powers, Harold, et al (August 1952). Geophysical Case History, Fredericktown Lead District, Missouri. *Mining Engineering*, 47(5) 517-42.

Queen, H.L. (1988). *Chronic Mercury Toxicity: New Hope Against an Epidemic Disease.* Colorado Springs, CO: Queen & Company Health Comm., Inc.

Queen H. L.(February 1989). *Heart Talk*, 8(1):5.

Queen, H.L. (May 1992). *Mercury-Free News*, 5(2):10.

Rabinowitz, M., Leviton, A. & Bellinger, D.(1983). Home Refinishing: Lead Paint & Infant Blood Lead Levels." *American Journal of Public Health*, 75:403-404.

San Francisco Examiner (15 December 1993). Dentist to Post Warnings on Mercury. A(1).

Schwartz, J. (1994). Societal Benefits of Reducing Lead Exposure. *Environmental Research*, 66:105-124.

Seeley, Rod & Graef, J.W. (1992). *Anatomy and Physiology, 2nd. Ed.* St Louis: Mosby.

Seth, T.D., Sircar, S. & Hasan, M.Z. (1973). Studies on Lead Extraction from Glazed Pottery Under Different Conditions. *Bulletin of Environmental Contamination and Toxicology*, 10:51-56.

Settle, D. M. & Patter, C.C. (1980) Lead in Albacore: Guide to Lead Pollution in Americans. *Science*, 207:1167-76.

Shannon, M.W. & Graef, J.W. (1982). Lead Intoxication in Infants. *Pediatrics*, 89(1)87-90.

Shiraki, J.H. & Nagashima, K. (1977). Essential Neuropathology of Alkylmercury Intoxications in Humans from the Acute to the Chronic Stage with Special Reference to Experimental Whole Body Autoradiographic Study Using Labeled Mercury Compounds. Neurotoxicology, pp. 247-260.

Staes, C. & Rinehart, R.(1995). *Does Residential Lead-Based Paint Hazard Control Work? A Review of the Scientific Evidence.* Columbia, MD: National Center for Lead-Safe Housing.

Sugarman, C.(November 1986). Plates That Bombed—Fiesta: Red Hot Tableware. *Harrowsmith* (Canada).

Swindell, E., Chancy, E., Brown, J.J. & Delaney, J.(1994). *Lead Level Changes in Children with Class III Lead Poisoning.*

Tavolato, B., Licandro, A.C. & Argentiero, V. (1980). Lead Polyneuropathy of Nonindustrial Origin. *Eur. Neurol.*, 19:273-76.

Townsend Letter for Doctors, 129:394. (April 1994). California Dentists Now Required to Inform Patients of Toxic Substances in Dental Fillings.

Trakhtenberg, I.M. (1974). The Micromercurialism Phenomenon in Mercury Handlers. *Chronic Effects of Mercury Organisms.* DHEW Publ. No. 74-473: NIH, pp. 109-134.

United Press Internat'l (12 Dec. 1983). 300 Get Lead Poisoned. *Seattle Times*, A(4).

U.S. Dept. Of Housing & Urban Development (1990). Comprehensive and Workable Plan for the Abatement of Lead-Based Paint In Privately Owned Housing. Report to Congress. Washington, D.C.

U.S. Dept. Of Housing & Urban Development, Office of Policy Development & Research. HUD 1991: The HUD (FHA) Lead-Based Paint Abatement Demonstration Project. Prepared by Dewberry & Davis. HC-5831, Washington, D.C.

U.S. Dept. Of Housing & Urban Development, Office of Lead-Based Pint Abatement and Prevention (June 1995). HUD 1995a: Lead-Based Paint Hazard Reduction and Financing Task Force. *Putting the Pieces Together: Controlling Lead Hazards in the Nation's Housing.*

U.S. Department of Housing and Urban Development, Office of Lead-Based Paint Abatement and Poisoning Prevention. HUD 1995b: Guidelines for the Evaluation and Control of Lead-Based Paint Hazards in Housing. HUD-1539-LBP, Washington, DC.

U.S. Department of Housing and Urban Development, Office of Lead-Based Paint Abatement and Poisoning Prevention. HUD

1995c: Lead-Based Paint Hazard Control Grant Program for Low-Income Private Housing. Washington, DC.

U.S. Department of Housing and Urban Development, Office of Lead-Based Paint Abatement and Prevention. HUD 1995d: Regulatory Impact Analysis of the Proposed Rule on Lead-Based Paint: Requirements for Notification, Evaluation, and Reduction of Lead-Based Paint Hazards in Federally-Owned Residential Property and Housing Receiving Federal Assistance. ICF Kaiser, Washington, DC.

U.S. Department of Housing and Urban Development, Office of Lead-Based Paint Abatement and Poisoning Prevention (March 1996) HUD 1996: Interim Report on the Evaluation of Controlling Hazards in Low-Income Housing. Prepared by the National Center for Lead-Safe Housing and the University of Cincinnati.

U.S. Department of Housing and Urban Development and the Environmental Protection. "HUD/EPA 1996: Joint Rule, Requirements for the Disclosure of Known Lead-Based Paint Hazards in Housing." *Federal Register*, March 6, 1996:   9063-9088.

U.S. Food and Drug Administration. Compliance Policy Guide 78117.07: Chapter 17—Food Related. Subject: Pottery (Ceramics); Imported and Domestic—Lead Contamination. Issuing Office: EDRO, Division of Field Regulatory Guidance. Authority of the Associate Commissioner for Regulatory Affairs, 10/1/80.

U.S. HEW, OSHA, Occupational Exposure to Lead, Final Standard. *Federal Register*,. November 14, 1978: 52953-53014.

Vimy, M.J., & Lorscheider, F.L.(August., 1985). Serial Measurements of Intra-Oral Air Mercury: Estimation of Daily Dose from Dental Amalgams. *Journal of Dental Research*, 64(8):1072-1075.

Virkkunen, M. Behavioural Changes Relating to Changes in Total Serum Cholesterol (1 February 1985) *JAMA*, 253(5).

Watson, W.S. (2 August 1980). Oral Absorption of Lead and Iron. *Lancet*, 2:236-37.

Weening, J.J. (1984). *Mercury Induced Immune Complex Glomerulopathy: An Experimental Study*. Van Dan Denergen.

Weden, R.P. (1984). *Poison in the Pot: The Legacy of Lead*. Southern Illinois University Press: IL.

Weisskopf, M. (December 1987). Lead Astray: The Poisoning of America. *Discover*.

Wessel, M.A. & Anderson, K. E. (1977). Our Children's Daily Lead. *American Scientist*, 65:942-46.

Whitehead, T.P. & Prior, A.P. (17 December 1960). Lead Poisoning from Home-Made Wine. *The Lancet*, 2:1343-34.

Whitney, H. (16 February 1994). The Biological Roller Coaster: Chronobiologists Study of the Body's Natural Rhythms. *Omni*, 16(6): 26.

Williams, M.K. (23 September 1972). Lead Poisoning: An Unusual Complication of Cystitis. *The Lancet*, 2:662-63.

World Health Organization (1977). Environmental Health Criteria 3. Geneva: World Health Organization.

Ziff, S. (1984) Silver Dental Fillings: The Toxic Bomb. *Aurora Press*. 8-9.